Walking with Shadows

Walking with Shadows

Josef Tarnowski

with

Raymond Raszkowski Ross

Glen Murray Publishing

Published in Scotland
by
Glen Murray Publishing
www.glenmurraypublishing.co.uk

ISBN
Paper-bound 978-0-9561758-0-9

A CIP catalogue record for this book
is available from the British Library.

Printed in Scotland
by
Bell & Bain Ltd

To the memory of all my friends and comrades
who fought and died for Poland; especially to those
who suffered and died at the hands of the Soviets
in the Gulags, the prisons and the torture cells;
and on the terrible transports to and from Siberia.

POLSKA RESTITUTA

Contents

Acknowledgements

Many people have encouraged me to write my story. Without that encouragement it would have remained in my memory for me alone.

I am grateful to all my friends and give special thanks to Jim Campbell in Edinburgh, to friends in the Rotary Club of Braids, to Robert Millar of Musselburgh and to David McCleary of Kilfauns Castle.

Introduction

Before I was born my country, Poland, had been occupied and partitioned for 130 years by three predatory powers: Russia, Austria and Prussia/Germany. Before that it had to withstand frequent invasions from all sides because of its position in east central Europe. As a result Poles had to learn how to survive and adapt or, to put it another way, they had to learn to adapt in order to survive as Poles.

It's not necessarily the fittest who survive but, as Darwin originally said, it's the most adaptable. This book is my odyssey of survival and change through a series of radically different and sometimes horrendous circumstances which came about with the outbreak of World War II.

September 1939 was one of the bleakest months in Polish history. On the first of September Hitler invaded our country from the West and though the Poles fought with bravery, and with much greater skill and perseverance than many western commentators (including historians) have given them credit for, they were eventually overwhelmed by technically superior forces. But we should not overestimate this superiority. German forces were concentrated on Poland, leaving the western flank undermanned and weak. Had France attacked Germany on September 3rd, Germany could have been beaten. But it was not to be. The French mind-set was defensive and they believed in their Maginot Line. Their alliance with Poland had always been about defending France, not Poland. So the game was up for us when, on the 17th of September, we were also stabbed in the back by the Soviet army suddenly sweeping in from the East.

By October our army had been crushed. No one had expected this. No one knew of the secret protocols in the Nazi-Soviet Pact signed in August that year. But as Poles we knew soon enough that the immediate purpose was to wipe Poland from the map once again: and this time to wipe millions of our people from the earth.

In relation to its size, population and the length of hostilities, Poland was to suffer far more than any other country during this war. Six million citizens, nearly twenty per cent of our population, would die, three million of them Polish Jews.

For those, like myself, who survived, life would never be the same again.

Everyone knows (and certainly should know) about the Holocaust. Annually tens of thousands of people visit Auschwitz and Hitler's other extermination camps, most of which were set up in Poland. These visits are important acts of remembrance and we should never forget the millions of Jewish people who were murdered in these camps, alongside Poles and other Slavs, gypsies, homosexuals, priests and pastors, resistance fighters, children and people with a physical or mental disability as well as those who opposed Nazism by thought or deed.

January 27th every year – the day of the liberation of Auschwitz – is Holocaust Memorial Day. Lest we forget.

Yet what we often do forget, at least in the 'West', are those millions of men, women and children condemned to slavery and to a slow death in Stalin's camps: the 'Gulags' which stretched across the USSR, from near enough Moscow itself right up and into the Arctic Circle and the Far East.

It's true that in these camps Stalin didn't use gas-chambers to murder his millions. He didn't need to. He had malnutrition, exhaustion, disease and, perhaps most fatal of all, the weather, to do that for him. What many barely realise, though it should never now be forgotten, is that in terms of grim statistics Stalin's murder count of the innocent far exceeded that of his Austrian counterpart.

Stalin killed his citizens by famine, by mass executions and deportations, by secret assassinations and official murders following show trials. And he killed them in the Gulags, ostensibly work (slave) camps or 're-education' centres.

In 1939, through the destruction and division of Poland, I became one of Stalin's citizens and soon enough one of his slaves and I would certainly have died in the Arctic Gulag had Hitler not inadvertently saved my life.

It is generally agreed now (see Anne Applebaum's *Gulag: a History of the Soviet Camps*), that some eighteen million souls passed through the Gulag system and that, of an estimated two million Polish civilians deported in 1939-40 to Arctic Russia, Siberia and Kazakhstan (see Norman Davies' *Heart of Europe: a Short History of Poland*), half were dead within a year.

Professor Applebaum reckons the total number of Soviet slaves (forced labourers), including six million exiles who were strictly speaking not Gulag inmates, to be nearly twenty-nine million. The authors of *The Black Book of Communism* calculate that Soviet rule caused the deaths of some twenty million people.

I am not a historian – I am an engineer – and I cannot compete with such scholarship. Nor would I wish to. But I do wish to honour the suffering of those millions through the telling of my own story, of my own experience as a young Pole condemned to the Soviet Gulag – to Vorkuta in the Arctic Circle – on Christmas Eve 1940.

This book is the story of my life. It is not simply a tale of the Gulag and though it tells of how that experience changed me forever, it also recalls my 'multi-cultural' Polish childhood, in what is now western Ukraine; it deals with my brief days in the Polish resistance (the *Armia Krajowa*), my arrest, torture and deportation, as well as my time in the slave camps until our forced liberation at the falling out of the thieves – when Hitler turned on his former comrade and eager helper Joseph Stalin and invaded the Soviet Union.

It also tells of our terrible journey then from the Arctic to Uzbekistan to form a new Polish army; thence (having by now contracted malaria) to Persia, the Middle East and Suez from where I criss-crossed half the world on a troop ship eager to avoid German U-boats before seeing the Promised Land for the first time when I landed at Gourock in Scotland.

In the Kingdom of Fife, in Manchester and in Peterborough, I was trained in the Parachute Brigade before being dropped into the bloody slaughter of Arnhem, an example of gross military incompetence where lions were certainly led by donkeys.

None of these things would I have survived without the people around me who gave me the energy and will to survive.

When the Allied betrayal of Poland at the Tehran, Yalta and Potsdam 'conferences' became public knowledge I was a 'soldier policeman' in occupied Germany faced with the choice of being 'repatriated' to Soviet occupied Poland (and a probable death sentence) or trying to start a new life in a country, Britain, which perhaps didn't really want me. I was a 'displaced person' in every sense of the word but I fought hard to adapt to my new environment, learning English well and quickly, and taking a qualification in electrical engineering at Portsmouth Technical College.

I was to work in electronics for thirty five years, my London-based career taking me to North America, South Africa and to almost every country in Western Europe, including, ironically enough, a three year stint in West Germany.

I was to visit a bleak and disturbing communist Poland twice for family reasons in 1971 and 1973; but I was also to make many trips there as a voluntary business consultant after *Solidarnosc* had finally seen off the communists. Or had it?

My continuing experiences in Poland are of a land in transition, but a land which still has bullet holes in the walls and in people's minds, a land struggling to know itself again and to heal itself, a land where the shadows walk with you still, but a land also of commitment to new things, to a new Europe and a new hope.

1
A Little Pole

I was born in the town of Maniewicze in Red Ruthenia in eastern Poland (now western Ukraine) on Sunday March 19th, 1922, St Josef's Day. This was doubly lucky, for I was born on a Sunday and my patron was one of the key figures of Christianity, the earthly father of Christ himself. Although I'm not now a church-going Christian, and although I have seen the worst as well as the best of times, I can safely say that my luck, whether divinely sourced or an earthly blessing, has always held. More or less.

One small childish downside, however, was that March 19th marked both my birthday and my 'name day'. In Polish tradition, a child receives presents on both his birthday and his name day, the feast of the particular saint he or she is named after.

As my birthday and name day coincided I did miss out on the double helping of presents! Maybe I was missing out in a small way; but in every other respect I was far from short-changed. Our family background was not without its shadows, but my own childhood was to prove idyllic.

My parents in Luck, November 1937

My father, Jan Tarnowski, came from Winnitza, later to become one of Hitler's HQ's for prosecuting the war against the USSR. My mother, Katarzyna Rozylo, hailed from the town of Ploskirow (now Ukrainian Chmielnicki).

In December 1917 the new Bolshevik government invaded the Ukraine, marking the start of the 'Russian' Civil War. During this dark period my father's first wife died from tuberculosis, leaving him with two sons (soon to become my half-brothers), Stanislaw and Jan.

During 1918, most of the Polish population in eastern Ukraine began evacuating west, towards

Josef, aged 6

Poland, and during this time my father and mother met. With Polish independence restored in 1919 (for the first time since 1795) they married and settled in Maniewicze on the border of the two Polish provinces of Kresy (which means 'at the edge') and Wolyn in 1920.

At first a clerk in the post office, Dad soon became post master in the town and we lived in a large wooden building, with a corrugated iron roof and a beautiful wooden veranda, which served as both post office and home.

Next door was a big orphanage (there were huge numbers of orphans following the First World War and the Bolshevik intervention in Ukraine and Poland). My parents were friendly with the manager of this orphanage, a Polish lady by the name of Zofia. Zofia had been working as a nurse in St Petersburg during the Revolution and, having witnessed the Bolshevik Terror first hand, she was vehemently anti-Soviet. Poles who lived in the Ukrainian/Russian border areas were usually very nationalistic. In Zofia's case you could say she had the red and white flag of Poland painted on her forehead.

It was she and my mother who taught me Polish history and patriotic songs from my infancy. The first nursery rhyme I remember learning from Zofia opened with the words:

Kto ty jestes?	Who are you?
Polak maly.	I'm a little Pole.
Jaki znak twoj?	What is your sign?
Orzel bialy.	The White Eagle

After centuries of partition, Poles felt it a duty to teach their children about their reclaimed identity and this fell to the women, the mothers, as the men were invariably working.

My parents were also friendly with the manager of a nearby country estate. Most of the estates in that part of Poland were owned by Lithuanian magnates and often managed by Poles while the peasantry was solidly Ukrainian. The old Greater Lithuania had stretched from the Baltic to the Black Sea and was part of the Polish Lithuanian Commonwealth which lasted from the thirteenth to the sixteenth century when the union of the Polish and Lithuanian crowns was effected under the Treaty of Lublin. The resulting Kingdom of Poland-Lithuania survived in one form or another until the final Polish partition of 1795.

The Wielkie Miedwierze estate was owned by Pan (Monsieur) Rdultowski who lived in Lithuania and who only visited occasionally, usually to collect rents and the like. The Polish estate manager was Ignacy Pankiewicz. Both Ignacy and Zofia were single and, with my mother in the role of match-maker, they married in 1925, when I was about three.

This was to prove brilliant for me. As I grew up, every summer was spent on the estate – Mum, me, Stanislaw and Jan, and – when he could get away from the post office – Dad.

In spite of the severe undernourishment I was to experience on my Siberian sojourn, along with a case of accidental self-poisoning, and subsequent bouts of malaria which began in Uzbekistan, I've always enjoyed good health and I think this must be rooted physically and perhaps even mentally in these summer months on the estate.

My summers were spent with horses: taking them to and from the fields, working them in the fields, riding them for pleasure. I often slept outside on hot summer nights under the trees in the huge orchards. I went fishing and swimming in the rivers and lakes or worked for hours on end in the dairy churning butter by hand. All the food was organic: good, fresh, wholesome, pesticide-free fruit, vegetables, meat and fish.

Round the *dworek* (the manor house), in which Zofia and Ignacy lived, was a beautiful flower garden where, every Sunday, Pani (Madame) Zofia would bring out a wind-up gramophone after lunch and play old Polish songs and operatic arias and we'd listen to Caruso and Paderewski and sing Polish and Ukrainian folk songs together.

We all knew, and sang, Ukrainian folk songs as well as Polish ones. The peasant women always returned from the fields singing and I learned Ukrainian by osmosis before I ever studied it at school.

At the end of every day the peasants would be paid in money and/or in kind: in grain, meat and potatoes. I used to watch them slaughtering the pigs. First, they were coshed on the head and then their throats were cut. The blood had to be kept for *kashanka*, similar to Scottish black pudding, only much better! Similarly, dozens of different types of delicious sausage could be made from good, fresh pig flesh – so delicious that when I met my first British sausage I thought it contained sawdust, albeit edible sawdust.

There was activity all the time on the estate with all the food being made and preserved for the winter: smoking the huge hams, pickling cucumbers in dill and pickling the cabbage for sauerkraut; all of this edible throughout the severe winters that could be twenty or thirty degrees below zero and when all the lakes and rivers were frozen over.

On Sundays family and friends would go picking blackberries or, later in the season, hunting for mushrooms. All the mushrooms we ate in those days were wild. We would have laughed at the idea of cultivated mushrooms. Wild and plentiful they were, but you had to be careful in the forests because of the boar which were equally wild and plentiful. They're aggressive creatures, especially when they have young, and it was not unknown for people to be killed by them. When attacked, you learned quickly to get up a tree as fast as you could.

The boar too were hunted in the autumn and winter and hunters would come from all over Europe. In the dense forests of Bielowieza, bordering on Bielorus, frequent pre-war visitors included Hitler's deputy Herman Goering and Mussolini's foreign minister (and son-in-law) Count Ciano. They were always accompanied by a bodyguard of soldiers. If the representative of the master race shot and missed his wild boar and it turned on him, as would happen, the soldiers were there to shoot the boar before it got its assailant! Just think how a small hairy pig might have had a significant effect on European history.

In these summer days and on these summer jaunts, we youngsters never went stealing fruit from orchards or any such thing that mischievous young lads might be expected to do; not because we were little angels, but because everything was so plentiful the thought never crossed our minds.

At home in the town I was also a bit of a train spotter (a habit that would come in handy when I joined the Home Army). Many trains from near and far passed through Maniewicze, often delivering the post to us which was then distributed from the railway station by horse-drawn vans. The roads were unmetalled and quickly turned to quagmires as winter approached. We did have pavements in places: wooden sidewalks or boardwalks of the type common in American 'wild west' towns.

The local church was also a wooden building with a separate, stone-built bell-tower. Bell-towers are often set apart from the actual church building in Poland, simply because the wooden structure of the church would never support a bell. Even now, with stone-built, brick or concrete churches, the tradition of the separate bell-tower is often maintained.

The summer idyll would end come September and school. In Maniewicze there was no middle (or 'secondary') school and I think, with our parents looking ahead, this was one of the reasons we moved to the larger town of Kowel when I was six.

In Kowel our house was close by a vodka distillery and the smell was overpowering. One of my school pals, Oskar's dad was the distillery manager; so we often played in the grounds of the distillery. I was inhaling vodka on a daily basis long before my palate began to appreciate its liquid subtleties.

My two closest pals were Leszek Ziolowski, whose father was a police inspector, and Jurek Potocki, whose family owned the only hotel in Kowel, the Hotel Polonia. We were 'the three musketeers' and we did everything together, inseparable in or out of school.

Leszek's family also had a grand piano and his mother was an excellent pianist; this was the beginning of my real love for classical music.

As musketeers we swore undying friendship, but we hadn't counted on the *gymnasium* exam. To get into the *gymnasium* (the high school), you had to pass this difficult exam. Only around ten per cent usually did. If you failed you went to the trade school (equivalent to the old secondary modern in the UK).

Luckily, I passed and got into the *gymnasium*. Leszek also passed, but he was packed off to the famous, private boarding school, the Krzemieniec Gymnasium, run by a religious order in the town of Krzemieniec, birth place and resting place of the great Romantic poet Julius Slowacki.

Jurek didn't pass. He went to trade school. That was the end of the 'musketeers'. But tragedy also struck. Within the year Jurek died of rheumatic fever. This was my first experience of death and it had a big effect on me.

I was now eleven years old. It was 1933 and we were on the move again, this time to Luck (pronounced 'Lootsk'), the capital of the district of Wolyn, where a huge new post office was being built and where dad had been appointed deputy postmaster.

2
A Multi-Cultural Society and the Threat of War

Looking back on my childhood in pre-war eastern Poland, I was brought up in what is now commonly called a 'multicultural' society; a pretty stable and largely happy society where different ethnic groups rubbed along reasonably well together. But it was not without its tensions.

In the old stratification the Ukrainian peasantry were bottom of the pile. The landowners remained largely the old Lithuanian aristocracy while the Poles occupied the middle stratum as administrators, police and army officers, and teachers.

The majority of the population in the countryside was Ukrainian, around 60 per cent. In towns like Luck, the majority population, around 90 per cent, was made up pretty equally of Ukrainians, Poles and Jews, with the final 10 per cent consisting of the Lithuanians, re-settled Germans and some Russians and Czechs (who ran the local brewery in Luck, for example). The Ukrainians, like the Jews, had their own places of worship and often their own schools and businesses, though most businesses were, in fact, Jewish.

The Ukrainians wanted independence and, I think, the Poles made a mistake in not giving them more autonomy in the new Poland. But that was the problem. This was a new state and the Poles were, not unnaturally, 'touchy' about Ukrainian autonomy. Poland had only recently (1918) been re-constituted – over a hundred years after it 'disappeared' in the final Partition of 1795. It had already withstood a Soviet invasion, inflicting the first defeat on Trotsky's Red Army, before stabilising its borders in 1922.

The 1918-22 military campaign was against not only the USSR but also against German uprisings in Silesia and Poznan and we were still thinking – as patriotic Poles – in terms of the old 1795 borders which stretched from the Baltic to the Black Sea.

The 1922 Treaty of Riga defined our eastern border, dividing the Ukraine into two parts: the Soviet controlled east and the Polish west, the old Polish land of Ruthenia where I was brought up. The lack of an independent Ukraine remained a bone of contention.

The Ukrainians formed an independence movement called 'RUCH' (which simply means 'movement'); but it was an illegal organisation and you could be jailed for membership. In the Second World War, a lot of Ukrainians (though by no means the majority) joined the *Wehrmacht,* believing Hitler would give them independence. They were eager to turn on the Soviets, especially in western Ukraine, that is, west of the River Dnieper which marked the old Polish Commonwealth border.

Eastern Ukraine was dominated by Russia and was always (as in many ways it still is) Russophile if not Russified, while western Ukraine was more influenced by Poland. The east had the coal, steel and uranium; the west was the grain basket.

We knew all about the famines in eastern (Soviet) Ukraine: the famine caused by the 'Civil War' and the Bolsheviks' scorched earth policy and the famine caused by Stalin's policy of forced collectivisation in the 1930s.

Mum got letters from her sisters in eastern Ukraine during the 30s describing how they were surviving on potato peelings and a lot of refugees fled into eastern Poland, despite Soviet border security, bringing with them stories about the terrible conditions. We always knew what the Soviets were about and what happened to their oppressed 'minorities'.

Although I was only 11 years old when we moved to Luck in 1933, the year Hitler came to power, I was very aware of politics. I had been born at a time when Polish borders had at last been defined after more than a century of partition and the Polish people could seemingly look forward to building their free, independent country. But the difficulties facing Poland were many. As a result of partition there were three different systems of administration, education, military training and technology which had to be reconciled. To complicate matters further, many different political parties had emerged but were unable to co-operate, each one offering its own solution to governing the new state.

The resulting political instability culminated in May 1926 in a military coup led by General Pilsudski, establishing the *Sanacja* regime. The coup had been opposed by General Sikorski, whose leading role in the defeat of the Red Army in 1920 had always been underplayed to Pilsudski's benefit. Sikorski had a lot of supporters in the army and he supported the democratic government of the day. Civil war was a real danger but, at the Poniatowski Bridge in 1926, Sikorski 'blinked' when Pilsudski refused to withdraw his troops. As the *Sanacja* came to power, Sikorski left Poland to become a lecturer at the French military academy.

At home, in school and in friends' houses we were all taught stories about Polish heroes and Polish history. Wars, for us, were about heroes and this had a tremendous effect on our young minds.

We all dreamed about becoming heroes if and when the call to arms came. We didn't realise then just how soon our 'opportunity' would arise.

Living in eastern Poland we naturally regarded the USSR as the big threat. Despite his doctrine of 'Socialism in One Country', we knew Stalin wanted to spread the 'Revolution' and, although we had non-aggression pacts with both Germany and the USSR, we probably trusted the former more than the latter.

Poland had started to re-arm with an eye to the East. A tank battalion was posted to Luck in 1934-5: French built tanks with French instructors. Poland had a mutual assistance pact with France, which was really more about defending France than about defending Polish borders. When Hitler occupied the Rhineland, Pilsudski suggested invading Germany (which might have been enough to call Hitler's bluff) but the French refused. France was disarming under the socialist government of Leon Blum, making it militarily very weak.

As the 1930s progressed we became more and more aware of the German threat. We watched the occupation of the Rhineland, the *Anschluss* and the occupation of the Sudetenland. But it was not until Hitler's troops entered Prague in March 1939 that we were certain we were next.

After Pilsudski's death in 1936, the *Sanacja* government shifted further to the right. It had no clear foreign policy and actually supported Germany over Sudetenland. Polish troops took the opportunity the crisis provided to smash into Czechoslovakia to annexe the disputed city of Trzecin (Cieszyn), through the middle of which the border had run. This was to become a big bone of contention between the Czechs and the Poles. It was short-sighted acquisitiveness on the part of the *Sanacja*, though a popular move among Poles in general. With hindsight the government claimed they had not wanted this industrial city to fall into German hands. This was an excuse. It was a wrong move politically which I think we were largely conditioned to support by clever government propaganda. In any case the Germans knew they would get it eventually.

Czechoslovakia had a mutual assistance pact with the Soviets but any Soviet intervention would have meant the Red Army crossing Polish territory and that the Polish government would never allow. As if to underline this, Warsaw staged huge military manoeuvres around Wolyn involving around 100,000 troops, about a third of the Polish army. At the end of the manoeuvres a great military parade took place in Luck reviewed by Marshall Rydz-Smigly, commander-in-chief of the Polish armed forces.

Watching this long parade, with all sorts of military hardware on display, we became convinced that we could deal with any invader. Well publicised in the press, the parade achieved its objective in convincing the nation about our battle-readiness, though it's doubtful if it convinced either the Germans or the Soviets.

With the annexation of Bohemia and Moravia, with Slovakia now a German puppet state, Poland was surrounded by Germany on three sides, the Soviet state closing the fourth. But when the Germans entered Prague in March 1939, the Hungarians also occupied part of Czechoslovakia, applauded by Poland. Close allies, they now shared a common border.

It was mainly across this border that the remnants of the Polish army would escape in the autumn of 1939 to continue the fight in France and the Middle East. It was also across this border (and the border into Romania) that the *Sanacja* escaped to disappear from most of the history books.

What happened to them was never much talked about in the Polish community in Britain during or after the war. When they got to France, much to their *chagrin*, they found that it was General Sikorski who was appointed Prime Minister and Commander-in-Chief. Not only that, but most of the *Sanacja* officers, with Sikorski's blessing, were interned as prisoners on the Isle of Man for the duration of the War. They were blamed for a policy that had been too pro-German until late 1938.

But in 1939 we still had confidence. In April that year Great Britain, which had the largest navy in the world, guaranteed our borders. France still had the biggest army in Europe (though, as we were to discover, its generals were locked in a defensive World War One mindset). No one, however, fully realised that we could not match German technology.

After the annexation of Czechoslovakia we were anxiously awaiting Hitler's next move. We fully expected that he would claim the Free City of Danzig (Gdansk) as part of the Reich – but he decided to embark on a much more ambitious ploy. Poland had already built a deep water port at Gdynia some miles north of Gdansk and outwith the 'Free City' territory. It had more modern port facilities and was handling a much bigger tonnage of goods than Gdansk/Danzig so the latter had more sentimental than economic value for Poland.

Hitler knew that demanding Gdansk alone might not be sufficient to provoke war with Poland. So he went one step further. He demanded a hundred metre wide strip of land across the 'Polish Corridor' to build a closed *autobahn* between the province of Brandenburg and German East Prussia. This, of course, would isolate Gdynia, cut off Poland from the Baltic, leaving it landlocked and at the mercy of Germany. Poland naturally refused.

Meanwhile, the Molotov-Ribbentrop pact was being drawn up with its secret protocol to partition Poland and wipe it once more from the map of Europe. Not only that, but both parties were to set about destroying the Polish nation, killing off the intelligentsia and rendering the remaining Poles slaves.

In March, Poland began mobilising in secret. My older colleagues were receiving call-up papers. This was allegedly for summer manoeuvres but we all suspected they were going to man defensive positions on our borders with Germany. War fever began to take a grip. We were all discussing what we would do when, not if, the war started. We all imagined ourselves as heroes.

Then, on August 24th, came news of the Nazi-Soviet Pact. Hitler was now accusing the Poles of persecuting Germans within its territory, seeking the excuse to invade. Provocations were staged including the infamous 'raid' at Gleiwitz (Gliwice) in Silesia where German convicts were dressed in Polish uniforms to 'attack' the German radio station. They had been falsely promised their freedom for this noble service to the Reich, but the attack went to plan and they were mown down by SS units. Hitler now had his excuse. Poland would be forced into war.

3
Poles and Jews

Relations between the Poles and the Jews had been good up until General Pilsudski's death in 1936. Then the disease of ultra-nationalism that Hitler introduced into Europe began to have an effect, with some Poles, a minority, imitating German racism.

Pilsudski was a military dictator and the *Sanacja* regime was a *junta* but Pilsudski had always followed a policy of toleration towards minorities. After his death the *junta* allowed excesses to take place. For example, it became an accepted principle that no university would take more than ten per cent of its student population from the Jewish community. This was never enshrined in Polish law, it was never a legal stipulation, but it was adhered to. Jewish university students were also made to sit apart from others. There was a lot of anti-Semitic propaganda in the Polish press in the late 1930s but it was never actually possible to pass any anti-Semitic laws.

Our cadet corps at the *gymnasium* was a paramilitary affair with a lot of square-bashing and some pistol and rifle practice which definitely drew on the Nazi Youth model. A lot of ideas were seeping through from Nazi Germany and Poles were happy to hear the tune of anti-Soviet propaganda playing in their ears, especially eastern Poles who saw Russia as the enemy. The influx of large numbers of Jews expelled from Germany exacerbated anti-Jewish feelings.

In schools Jewish pupils were not made to sit separately. I had many Jewish friends like Marek Feldman who was the star pupil in my class and whom I was to meet again on my way out of Siberia and, indeed, in more recent years. Very clever and fairly well-off, Marek's family were shop-owners. A lovely chap, he wasn't entirely orthodox as he'd secretly relish the occasional taste of a good Polish pork sausage!

There's an old Polish saying that if you want a feast then go to the home of a Jew. The Jews were culinary masters (or mistresses), especially at cooking fish, and they had a great influence on the local cuisine. To this day I've been addicted to *matzos* and to dishes like pike or carp in aspic jelly.

And like the food, Jewish music was everywhere in my youth.

You could hear *klezmer*, really good *klezmer* music, in your friends' homes, in hotels and restaurants and in the streets.

There were also a good number of Karaim. Tradition (or myth) had it that the Karaim were Jews who had been expelled from Palestine by their orthodox brethren because they didn't accept the Talmud. Becoming wanderers, they drifted into Syria and through Turkey up to the Crimea. The Tartars accepted them because they were literate (unlike their hosts) and good accountants who could tally the riches the Tartars gathered from their raids into Poland.

But it seems more likely that the Crimean Karaim are people of Turkic (possibly Khazar) descent who adopted Karaism, which does not recognise Talmudic thought. By the twentieth century the three centres of Karaim settlement were in Luck and Lwow in eastern Poland and in Troki, near the Lithuanian capital of Vilnius (the Polish centre of Wilno).

The Karaim had their own Judaic religion, their own customs and their own dress. They were never consistently persecuted by the Nazis because they were not seen as Semitic. Their position was confused: although some Karaim collaborated with the Nazis there were many more who fought against them.

A thriving society of several thousand before the War, there are now only a few hundred in Poland (and possibly less than 5,000 in the entire world). Two of my Karaim friends from Luck are still alive today: Marek Mordkowicz who lives in Bytom and Boguslaw Abkowicz, who lives in Wroclaw. One of Boguslaw's daughters has written a history of the Karaim.

I feel infinitely sad that this culture has all but disappeared from Poland and that so little is even known about them in the wider world. Their temples in Lwow and Luck were destroyed but the one in Troki survives.

It was on a school visit to Troki in 1937 that I first tasted *halva* which, I think, must have been introduced by the Karaim. It was a unique delicacy I'd never come across before. It still is a great delicacy to me; and the taste of it always reminds me of the Karaim.

4
School-time to War-time

Josef (second row on right, leaning on friend's shoulders)

Our *gymnasium* was one of the best in Poland and I think I was considered a fairly clever young man who didn't work too hard. More interested in sports than scholarly pursuits, I became a red-hot Arsenal fan. Arsenal were *the* team, the Manchester United or Real Madrid of their day, managed by Herbert Chapman and with seven players in the English national team.

I played a lot of football and did a lot of rowing (we had two rowing clubs on the River Styr at Luck) and swimming, sometimes for miles. In the winter we played ice-hockey or skated on the river, though you had to be careful to avoid places where the ice was being cut for storage in the ice cellars where, wrapped in straw, it was used to preserve food during the summer months. The Styr was frozen from December to March and another favourite pastime was ice fishing: cutting a hole in the ice and spearing the fish as they came up looking for food. The easiest fishing possible!

One bizarre annual event was a kind of ritual 'ice-dipping'. which the Orthodox Ukrainians (who seemed to be split equally between the Russian and Greek traditions) indulged in on the Feast of the Epiphany. Based on the old Julian calendar, this was celebrated two weeks after the Catholic celebration. On this January day, when the air temperature could be as low as minus 35 degrees centigrade, they cut holes in the ice and the men jumped in! I think it was something to do with baptismal rituals (imitating Christ's baptism in the river Jordan): the Ukrainians always seemed to baptise in the river rather than in the church.

The Styr flowed into the Pripet Marshlands north of Luck, a huge wetland, practically impenetrable in parts, about the size of Scotland. A short ride by horse-drawn carriage, we used to go there to collect hazel nuts, pick blackberries and *poziomki*, a type of wild strawberry, hunt for mushrooms and fish for carp, pike and *sandacz*. Carefully placed logs of wood made parts of the marshes passable on foot but never accessible to heavier transport. This made it the best of hiding places for partisans throughout the war.

When the Germans invaded the USSR they had to go both north and south of the Pripet Marshes, there was no way through this natural barrier. But thousands of partisans could hide out there living off wild animals, fish and fruits of the forest.

There were five million people in the Polish resistance (the Home Army). How did the majority survive? That forty per cent of Poland is forest or wild wetland is part of the answer.

At the *gymnasium* I was more interested in the humanities (and sport) than the sciences and I thought I wanted to become a lawyer. I studied Polish, Latin, French, Ukrainian, Russian, world literature, history, geography, maths, physics, chemistry, biology, music and religion.

Music was compulsory. Poland is basically a musical country and we had monthly classical concerts at the school, often given by prominent musicians who came from all over Poland. I took up the violin for a short time but when my pals met me in the street and, looking at my violin case, asked me why I was carrying a coffin, I said 'That's it! I'm not learning it any more!'

The *gymnasium* was a real learning community. If any pupil was struggling with a subject it was his/her fellow pupils' duty to help them at home (yours or theirs) with their studies. That was the way it was. Everyone wanted the whole class to get good grades and

June 1938. Josef (seated) with his friend, Przemek Halacinski, whose father was executed at Katyn

the pride of the school was always at stake. We were all enjoying our new-found freedom in an independent Poland and, as young people, we wanted to build a strong and prosperous country. The spirit of co-operation was everywhere.

I remember a school dance when I was about fourteen or fifteen with the girls all dancing and the boys grouped awkwardly against the walls. This just wasn't good enough, so the class committee (the pupils themselves) organised a Sunday afternoon dancing class at the school where the girls taught the boys how to dance properly.

One subject you were not allowed to fail was religion. When I was about sixteen, the school chaplain Father Galezowski was giving a lesson on Christian ethics when a classmate asked him 'Can you buy your way into Heaven?' The priest was shocked, but asked him to explain himself. 'Well, supposing two people die at the same time and are both sent to Purgatory for ten years, but one is rich and one is poor', he said. 'The rich man's family can pay for lots of Masses to be said which will get him into Heaven quicker, while the poor man's family can't afford to do that. So, surely you can buy your way into Heaven.' The priest had no answer. So he threw him out of class and told him to report to the head teacher, the *gymnasium* director!

My parents were very religious and I have to say Father Galezowski came up trumps after my arrest. He supported my mother through the darkest hours, giving her hope and spiritual comfort.

As a boy I was conventionally religious, never questioning anything. Every Sunday at 9am you had to line up at school to be registered present before going to Mass together. But we had a system. We took a stop watch and went round the different Masses being celebrated in the various chapels of the cathedral, timing them. When we found a priest who could get through a Mass in twenty minutes we knew we'd found our man!

5
Poland Invaded: Resistance and Collaboration

As young patriots we were almost looking forward to the approaching war. Our hearts sank when it was announced that Russia and Germany had concluded a pact. How much further our hearts would have sunk if we'd known about the secret protocol it contained.

On the 30th of August a general mobilisation was declared which, if it had been held to, would have readied our forces better for the invasion which came on the 1st September. But Britain held it back. Ambassador Henderson was negotiating in Berlin and London didn't want to offend the Fuhrer.

By the next day it was obvious that negotiations were pointless. Warsaw mobilised but it was far too late to shift thousands of troops to key border areas. On 1st of September I watched a German reconnaissance plane fly over Luck. As a member of the *gymnasium* cadet corps I had been stationed on the main bridge to look out for German paratroopers.

We still had great hopes. What Polish heart does not? We held the Germans for three days before they broke through the Polish Corridor cutting it off from the rest of the country. This allowed General Heinz Guderian's panzers from East Prussia to attack the Polish army from the rear. By September 12th the Germans were at Brest-Litovsk on the river Bug. Five days later, the Red Army invaded. Poland's fate was sealed.

The Soviet forces immediately occupied Eastern Poland (what today is known as Western Ukraine). Our life changed completely. Luck, a town with a population of some 33,000 souls, was now to experience 'the joy' of living in the Soviet State. Almost overnight all food and all goods disappeared from the shops. Everything was confiscated and 'exported' to other parts of the Soviet Union without any compensation to the shopkeepers. There was no such thing as 'private property' now. Everything belonged to the State.

Every day everyone had to queue for hours on end for even the most essential foods like bread, milk, sugar and salt. Whole families, including children, had to queue. Or starve.

The only items I can remember remaining on the shelves were

matches. Manufactured goods, including even children's clothes, were totally unavailable and, with winter approaching, this posed a serious problem for families. Improvisation was the order of the day. The women turned their hands to tailoring, but shoes presented the biggest problem.

The entire Soviet manufacturing industry seemed geared to the needs of the military. The only plan the Soviets had for Eastern Poland was to invade and suppress all opposition with maximum force. There was no economic strategy beyond immediate theft and the imposition of a centralised state economy which led to chronic shortages, especially of consumer goods. People quickly learned to survive by bartering their possessions for food.

At the age of seventeen I was already a schoolboy 'soldier'. We had all been shocked when the Soviets suddenly stabbed us in the back just as we had been shocked by the collapse of the Polish Army. Many young men – and many young women – felt it was their duty to enlist in the *Armia Krajowa* (AK), the 'Home Army'.

I joined in the winter of 1939-40 after I witnessed the first deportations of families to Kazakhstan. During the September invasion itself all Polish Army reserve officers, policemen, local administrators, aristocracy and other local notables had been arrested. In the severe winter, with deep snow and temperatures of minus 35 degrees, the invaders came for their families.

Wakened at midnight and given one hour to dress and collect what they could carry, these women and children were taken to the railway station and loaded onto cattle trucks similar to those used by the Nazis to transport Jews, Poles and others to their concentration and death camps. With one stove in the middle of each truck, very little wood to burn and only meagre daily rations, these families were to endure a journey of thousands of miles. This exile was their punishment for the 'sins' of their fathers and their brothers. It was their punishment for being Polish. For me, watching them being loaded onto the cattle wagons was the trigger. I knew I had to do something.

Most of the reserve officers were executed at Katyn, or at Kharkov and Ostashkow, as we were soon to learn. It became common knowledge because the pre-war Polish government had a pretty effective spy network in the USSR and news was still coming back particularly from Belarus and the Ukraine and because the Soviets released some four hundred officers who had signed papers to the effect that Poland should be a willing ally of the USSR. They returned, along with NCOs and ordinary reservists who had escaped the bullet, and brought the news.

The Soviet authorities couldn't organise the supply of food and essential goods to the population but they could organise mass deportations (and executions). Altogether, some 1.5 million people were deported from the occupied territories of Poland, in effect, from a country which was not even at war with the Soviet Union. Of those, 250,000 were to be released to the Polish Army when Hitler invaded the USSR in June 1941 and a further 250,000 were to be released to return to Poland after Stalin's death in 1953. The rest perished, with the exception of a few who chose to remain in Kazakhstan or some other remote area of the USSR.

What happened to their homes? They became state property and were immediately allocated to NKVD and Soviet Army officers and to administrators shipped in from the East.

There were other delights of occupation. Gatherings of more than three people were forbidden. Local services collapsed. There was no guarantee of running water or of sanitation and waste disposal. People had to cope as best they could. Russian and Ukrainian became official languages overnight, to be used in all offices and schools. Letters were censored and there was no communication outwith the USSR.

Churches were shut and church property confiscated. Luck Cathedral became the 'Museum of Atheism'. Priests (Catholic or Orthodox) were arrested and jailed or transported.

Informer networks were set up. Some young people, particularly from the Ukrainian community, were recruited into the local militia. Poles were banned. It was 'divide and rule' as the Soviet authorities promoted strife where there had been reasonably good intercultural relations.

Private transport was confiscated. Fuel was unavailable. You had to pay exorbitant prices for wood or go into the forests yourself to collect and carry it home by hand.

Our main job in the AK in those early days was to gather intelligence, to collect and store as many weapons as we could get our hands on and to keep our heads down until further orders came from the Polish Government in exile in London. The town was awash with guns, all ex-Polish Army weapons. I had managed to collect a couple of machine guns as well as some rifles and revolvers. Wolyn was full of forests and our efforts soon had those forests bristling with hidden guns.

We'd all had some military training at school. I knew one end of a rifle from the other though it's maybe fortunate we weren't ordered into any military action just then! Most young men were keen to be involved and this must have been apparent to the Soviet authorities. After all, the Poles were great patriots and had never taken kindly to Russian rule, no matter what political guise it came under. You were almost expected to join the underground. It was simply what one did.

So it was no surprise that almost every young man in Luck between the ages of sixteen and twenty-one was arrested by the NKVD (later to be known as the KGB). Arrests started immediately the Soviets came in and continued right up to the German invasion in June 1941. The strategy was simple. The communist authorities wanted to infiltrate us (and the Ukrainian nationalists) with informers. They wanted to find the opposition and neutralise it, taking us out one by one.

Officially, of course, the Soviets were our liberators, though what exactly they had liberated us from we never knew. All we knew was that one day we were Poles living in Poland, the next we were Soviet citizens living in Ukraine. We had been incorporated into the Soviet Union.

Naturally the Soviet forces behaved like occupiers, not liberators. There were two sailing clubs in the town, for example, which were immediately 'reserved' for the NKVD and for Soviet army officers. Between them they also occupied the best buildings, hotels and restaurants. It was absolutely an *apartheid* system. They enjoyed their 'classless' existence; we were there to serve.

As these places were out of bounds to Poles, any Pole seen leaving or entering these premises immediately became suspect. I would say that out of a population of around 33,000 in Luck some two or three hundred did collaborate to some extent or other.

A person arrested by the NKVD might be released if he showed a tendency towards collaborating; that is, towards informing. It took us a few months to realise we had to keep away from those released. Of course, if these people didn't 'deliver' they were soon rearrested and imprisoned. However, I don't recall any suspected collaborators or informers being shot or punished by the AK in that first year of occupation.

The German SS were also in residence for about six weeks in the winter of 1939-40. They too had their own headquarters in the town and their own restaurants and hotels. The image of SS officers wandering freely around a Soviet occupied town may seem strange to people now but such was the case. In the early days of the war the Soviets themselves proved to be enthusiastic Nazi collaborators.

Luck had a fairly sizeable German population, an enclave of about two thousand who had settled there during World War One. They lived mainly in their own district and they rubbed along well enough with local Poles and Ukrainians, though we didn't really mix that much, mostly because of their 'attitude'.

From 1933, the year Hitler came to power, during every summer vacation, young Germans were invited to 'holiday camps' in the Reich. There was nothing unusual in this. We Poles traditionally had our own summer excursions and camps. What we didn't know then was that not only were a lot of these young people being propagandised but some were being trained as pilots. In September 1939 some of these holidaymakers returned as Luftwaffe pilots to help bomb Luck. We hadn't realised this training was going on. Some of our erstwhile neighbours were maybe even trained up for the SS. Who knows?

Under the terms of the Nazi-Soviet Pact the Volga Germans and the little enclave of Wolyn Germans in and around Luck, were to be 'repatriated' to Germany. The Soviets had invited the SS in to oversee the business. In fact, these Germans weren't repatriated to Germany *per se*, but to the western areas of Nazi occupied Poland which had been incorporated into the Reich.

On a daily basis these local Germans reported to the SS headquarters in Luck to arrange their return 'home'. I saw no evidence of actual fraternisation between the SS and the NKVD. They didn't parade together or openly socialise. But the SS presence indicated just how closely the Soviets were willing to work with their 'ideological enemies' during the first two years of the war.

As AK members we had made it our business to do a bit of trainspotting. Every day we saw heavy goods trains passing through from the Ukraine heading for Nazi Germany openly hauling loads of iron ore and coal and it wasn't difficult to spot those carrying oil or petrol in their cylindrical tankers. Tarpaulin-covered wagons we guessed to be carrying armaments and ammunition as well as grain and other foodstuffs. It was obvious to us in Wolyn that the USSR was supplying Germany with stock and *materiel* for war with France and Britain.

6
Arrest

On a hot summer's day in July 1940 the new director of the *gymnasium*, a Ukrainian we assumed to be NKVD (how else would he be director?), called me in to see him. 'We're going to have an excursion for the younger pupils' he told me. 'But we need some older pupils to help out. Would you be willing to lend a hand?'

Without thinking, I said I would. I was maybe a bit naïve: I entertained no immediate suspicions. It was the holidays after all and excursions *were* being organised. 'I can't give you all the details', he said. 'But there's a man at the town hall who can. You've to go and see him.' As it was the holidays this also made sense. The town administration organised such things.

It was a beautiful day. I was wearing a short-sleeved shirt and tennis shoes. The town hall was fifteen minutes walk, up past the army barracks and the bank. As I made my way up that road, watching the sun glistening on the cobbles ('cats' heads' we called them) I was thinking about what kind of camp it would be and what I'd do there. It was a challenge. I was really looking forward to it. The countryside around Luck was rich with meadows, forests and rivers. To get out there for a week would be heavenly.

Summer camps are a Polish tradition, giving parents a break during the long holidays. It was a tradition I was glad to see the Soviets had adopted. On that lovely summer's afternoon I was on my way to a camp. What I didn't realise was that it would take around nine months to get there, and that the camp I was going to would not be in beautiful Wolyn, but in the Arctic Circle.

I had said to my father a few weeks before: 'Dad, watch out. I might be arrested.' I felt the net was closing in. I expected it would come with a midnight knock, the usual NKVD tactic we all knew about and feared. But they were refining their tactics, as I was about to learn.

At the town hall I gave the receptionist my name and explained why I was there. She directed me into an adjacent office to wait. The room was bare except for a table and some shelves lined with the collected works of Lenin and Stalin.

At this point she probably picked up the phone, as no doubt instructed, and called NKVD headquarters. The man arrived within minutes. It was obvious he was NKVD. You could smell it. He wore a typical Russian cloth cap with a button on the top and the peak turned up. For a whole ten minutes he neither looked at me nor spoke. But his silence was eloquent. As he examined the books on the shelf, as he picked one out to flick through its pages, as he replaced it with the air of one pre-occupied, he was letting me know he was the authority; and that I was in his power.

'You know what?' he said in Russian. 'This isn't a nice place to discuss the details of the excursion. I think we should go to my office. It's much more comfortable and it's only a five minute drive.'

He grabbed my arm, led me quickly from the office, a few paces across the front hall and down three steps to a waiting black car.

I understood Russian. I also understood for sure that he was NKVD when he pulled a revolver on me and bundled me into the back of the car. Then a blanket was thrown over my head which was shoved to the floor and the gun pressed against it.

'Don't move' said the NKVD man.

I didn't. But the car certainly did. It was twisting and turning through the streets of Luck, zigzagging, doubling back, going in circles, all to disorientate or frighten me. This was certainly a strange and different excursion to the one promised by the director. But I knew the town too well. When we stopped after half-an-hour I knew exactly where we were. We'd arrived at the old regional administration offices, now NKVD headquarters.

My head still under the blanket, I was bundled inside. Once in his 'office' the blanket was pulled off and the man put his gun away.

'You know where you are?' he asked; more of a statement than a question. 'Well, there's something else you should know. There are only two ways out of here. You can walk or you can go feet first. I want weapons and I want names!' Choose!

Then he paused. 'Give me names, tell me what I want to know, and I'll let you go free.'

Later on, I was to realise that the competing totalitarian powers in Europe, Germany and Russia, operated two different methods of 'population control', particularly in Poland. The Germans' primary method was brute force or absolute terror, as exemplified by what happened in Bydgoszcz in western Poland. If I had been in German hands I would have been shot on the spot or hanged from a street-lamp or telegraph pole with a placard round my neck.

The Russians could be just as brutal but, in what might pass for Soviet subtlety, they took a longer view. Their attitude could be summed up as: why shoot them when we can work them to death?

I was lonely and frightened, but at least I knew I was now sharing the same fate as friends and colleagues who had been arrested before me.

7
Initial Interrogation

I was interrogated for four days and nights. Non-stop. No sleep. Standing sixteen hours at a time. It was the usual method of 'bad cop - good cop', only the 'bad cop' beat you constantly, either with his fists or with a length of solid rubber, a two-foot long truncheon. The other one simply watched and picked you up again and again after you'd been beaten to the ground.

The mental violence was also intense and I became totally confused. I kept saying that I knew nothing but that wasn't going to keep them at bay for long. I was also worried because my parents wouldn't know where I was. They might guess but they wouldn't be officially informed. And they might be implicated.

The interrogators worked in two-hour shifts. Always the same good guy - bad guy routine. The same beatings. The same questions. 'Why don't you just admit you're a member of a counter-revolutionary organisation?' 'How many weapons have you hidden?' 'Where are they?' 'What is your position in the organisation?' 'Why make this harder than it has to be?' 'Why make it harder on yourself?' 'Who are your friends?' 'Where are they?' 'What are their names?' 'What do you talk about with them?'

I tried to get away with giving only the names of people I knew who had already been arrested, who had fled into German occupied territory or who had simply and cruelly disappeared trying to cross that terrible border. But they weren't swallowing this and they were wearing me down rapidly.

I was told 'Write your life story from the day you were born right up to today.' I was given pen and paper and left alone to write my life story, trying to mention as few people as possible and trying not to incriminate myself or any one else. I was then put into another, bare room and left for two hours.

Brought back again I was told 'Write your life story'.

'I already have.'

'Do it again.'

I did. Then it was two more hours of isolation before I was dragged back into the interrogation room.

'You're a liar, aren't you? Look! These two versions don't even match. There are differences here and here and there. You're making things up! You're a stupid liar! Write it again.'

More beatings. More questions. More writings. This happened dozens of times throughout those first days, always focusing on the inconsistencies in what I wrote, whether they were major ones or minor. First and foremost it was about breaking you down. 'Truth' didn't come into it. I was totally exhausted and utterly confused.

'You're a worthless counter-revolutionary! You're not even worth the price of a bullet!' I was told.

Their *prima facie* evidence for my being a counter revolutionary was that I had once parked my bum on a portrait of Stalin. This had happened during one of the interminable public and youth meetings they held when they first came into Luck, meetings to glorify Stalin and the USSR and to educate us all in the wonders of Soviet communism. During these meetings all the students had to hold up a placard of some great Soviet hero or other, in my case smiling Uncle Joe himself.

As this particular meeting went on and on I just got tired and put my placard down to sit on it. It wasn't actually a counter-revolutionary gesture on my part, just tiredness and, yes, boredom. And this had happened months before. They'd had their eye on me ever since.

The isolation cell I was held in between interrogations was one of twenty recently constructed 'dog boxes' designed, one presumes, as 'transfer cells'. They were wooden cells about the size of a wardrobe with no windows. Half a dozen holes were bored into the door for air and there was a simple wooden bench. You were in isolation but you could at least communicate in whispers with the two prisoners immediately to your left and right. News travelled fast around the 'dog boxes'. New prisoners could tell you what was happening on the outside and you could ask others about your friends already inside.

At one point I was moved to another small cell. It was a single cell, but it had another occupant, a friend. We were left there a couple of hours together and wisely kept our conversation to the everyday and the banal. Undoubtedly, someone would have been listening in with some kind of device or other. We hadn't actually worked this out, we were just being careful. But the guards gave it away when they opened the cell door and faked surprise that two prisoners had been locked in together and then dragged me back out to my own cell. They thought they were being clever but the fakery was obvious.

Anyway, we'd spent most of the time playing tiddlywinks with the buttons from our trousers. They'd been cut off, you see, in case we tried to run away. Superior Soviet technology at work – if you run, we have ways of making your trousers fall down!

After four days of non-stop brutal interrogation you'd sign anything just to get a break from it. The object of the exercise was to get you to sign the confession they wanted, whether it was the truth or not. I signed. I admitted to being a member of a counter-revolutionary organisation and was immediately transferred to the nearby St Brigitta's Convent which they had converted into a prison.

8
St Brigitta's Soviet Prison

Having now been officially 'confessed', at least initially, my parents were informed of my arrest and my mother was able to send me some warmer clothes. Most of my friends were already in St Brigitta's and it would be safe to say that, Luck being a small town, between us we knew most of the thousand prisoners or so held here at any given time. About half the prisoners were Polish, half Ukrainian nationalists, members of RUCH. We all got along as fellow prisoners.

The holding cell in St Brigitta's was bare. The lights were kept on all night. There were no bunks of any kind and you slept on the concrete floor on top of your coat, if you were lucky enough to have one. There were twenty of us in the cell, which was just big enough to allow those not being dragged out for night-time interrogation to stretch out for what sleep might come our way.

The toilet facilities consisted of one bucket, with a lid, which was slopped out every morning. The smell was terrible. Once a day we got a piece of black bread to eat and twice a day a bowl of watery beetroot soup. We were always hungry, but everyone had to sacrifice a little piece of their black bread – to make chess pieces.

Chess, you could say, is a Polish passion (something we share with the Russians) and in the jail it was the only recreation we had, except talking about food or girls. The heavy black bread, full of water, was easily shaped into a particular chess piece before being left to dry. The game occupied our minds and helped to keep our fears at bay. So you continued to play even when you were sick fed up with it.

Although chess wasn't allowed there were three or four sets on the go at any given time. We always had a look-out with his eye to the Judas hole in the door to warn us if a guard looked in. Chess sets were confiscated regularly, which meant sacrificing more precious bread to make new ones.

The first person I met in the holding cell was a little Jewish fellow who said to me 'Joe, I'm a communist! When the Soviets entered Luck I greeted them with flowers! So, what did they do? They arrested me for being a Polish communist! 'We don't need Polish communists', they said. 'We have our own!' Such was the nature of Stalinist 'internationalism', the nature of the beast.

About ten o'clock at night we'd all settle down and try to get some sleep. Around midnight the cell door would open more often than not and three or four names would be shouted. If you were shouted you'd be taken back to a 'dog box' until they were ready for you. Then your interrogation would begin again. The same brutality. The same questions. At the end of each torture session you were forced to sign another statement. Then, about six in the morning, you'd be dragged back to the holding cell and left in 'peace' for maybe three or four days.

Then it was your turn again. Your last written statement would be compared with previous ones, inconsistencies picked over, vicious beatings resumed because you were a liar, you couldn't be trusted, you were trying to wriggle out of things. This carried on for two months as they tried to build up profiles of you, your family and your friends. Their intention was to implicate as many of your family and friends as they could.

The beatings, in fact, became more vicious as they forced you to stay on your knees for hours. Finally, when they had sucked everything out of you that they could, when they'd squeezed the lemon dry, you were given a document stating what you were accused of which you were forced to read and sign. This was your final confession, your admission of guilt.

Dragged back to the holding cell, another waiting game began. The next time they came for you, your name would be on a list of those going to 'trial'. We all knew how these lists operated. If your name was near the top your 'crimes' were considered the most serious. The sentence would most likely be death.

When they came for me there were eight names on the list.

Mine was first.

9
Trial and Error

Stalin used to have people shot out of hand when it suited; but one of the strange things about Soviet barbarism was that once they had decided to put you on trial, albeit a show trial, they observed Soviet law and procedures.

I had been arrested (as I found out) under Article 58 of the Soviet Codex. What I didn't know then was that under Article 58 you were not meant to survive. They were lining up a firing squad even before my interrogation, never mind my trial.

It took five months to put me on show, on Christmas Eve 1940. Although the trial was held *in camera* the court was full. But only of young NKVD officers there to learn, no doubt, how to do it properly. Outside, my parents and friends were huddled, waiting in the freezing cold and the snow.

The presiding judge was a Russian, shipped in to aid the cause of Soviet civilisation in Poland. He was accompanied on the bench by a local 'judge' (a Ukrainian communist) and one other (also a party official, no doubt). The eight of us, each of whom had his own separate 'defence' lawyer, stood in the dock.

The prosecutor first read out the charges against me, declaring: 'Josef Tarnowski is guilty not only of being a member of a counter-revolutionary organisation but of being one of its chief organising leaders. He is a wolf in sheep's clothing. I demand the highest possible sentence – execution!'

I was shocked beyond measure. I hadn't expected this. In spite of being top of their list I knew I was pretty small fry in the greater scheme of things. I was numbed. When all the charges had been read out against each of us the court was adjourned. No evidence was offered. No witnesses called. Our 'confessions' were all that was needed.

We were then left in the courtroom for two long hours, our lawyers milling around, coming and going, occasionally addressing us. Did the judges deliberate our fates during those two hours? Consider our 'confessions' in detail? Or drink coffee and read the papers? I have no idea. But we were assured by our lawyers this was normal procedure.

We discussed our predicaments, falling into periodic silences. As I sat brooding on my fate my defence counsel came over to me with some cigarettes from my parents.

'Don't worry', he said, handing me the cigarettes. 'They won't shoot you.'

'Why not?' I asked.

'Just watch me,' he said.

Now, my defence counsel was one of them, a communist, a trained Soviet lawyer appointed by the court. I wasn't exactly brimming with confidence. He was, after all, just part of the show. So, I wasn't surprised when he stood up in court and said 'My client is as guilty as hell!' But then he added 'He's as guilty as hell but you can't shoot him. You can't shoot him because he was not yet eighteen years old when the offences were committed.'

That's all he said and he sat down.

They had me where they wanted me, 'guilty as hell', but they couldn't get rid of me the way they wanted. By shooting me. Under Soviet law I could not be executed. This called for another adjournment to decide my fate. I felt enormous relief. I was now convinced that I would survive no matter what. My luck was holding.

I was sentenced to ten years in the Gulag, my transportation order signed by Nikita Khrushchev as Secretary General of the Communist Party of the Ukraine. I was to spend ten years 'in a remote part of the USSR for re-education' as it was put. It's the little details like this that showed you how the system worked. Officially, the Gulag camps were being described as 're-education' camps, but if you happened to die during this re-education process – tough! Who cared?

The court ordered the confiscation of all my private property, an irrelevant procedure as I didn't have any. I was sentenced never to come back to where my 'crime' had been committed, in short, never to return home. And my Soviet 'passport' was taken from me because I didn't deserve to have one. That was no hardship either as I'd never wanted one in the first place but, in effect, what they were doing was stripping me of citizenship and of identity. I was now state property to be done with as they saw fit.

I was elated just to be alive, not to have been shot, and I was relieved just to have got ten years. Everyone knew, of course, that few, if any, could survive ten years in Siberia. But we also knew that sooner or later Germany would attack the Soviet Union. It was a matter of surviving till then. We had talked about this endlessly since the occupation had begun. As Poles we knew the Molotov-Ribbentrop Pact was only a transition, that the real fight for Europe was still to come.

Hitler was buying time in which to destroy the Western Allies before turning on the USSR, and Stalin was pouring Russian oil into

Germany through Poland to support the German drive west. He wanted Germany to launch an assault on Britain. He wanted both the Axis and Allied powers punch-drunk and on the ropes, making them easy meat for the Red Army which could then advance communism to the shores of the Atlantic.

10
Visions of Siberia

I was put in a larger holding cell for those awaiting transportation, about one hundred and twenty souls in all. Here I was re-united with some of my school colleagues who had been arrested before me, some of whom I'd mistakenly thought had made it over the River Bug into German territory. It's ironic looking back now, but a lot of us did think we'd be better off in German occupied Poland at the time.

Life in the holding cell was claustrophobic and extremely boring. Every couple of days, if the snow wasn't too deep, we got a half-hour walk round the perimeter of the prison, eyes down, no contact, no talking.

In the cell itself we ran occasional classes on history, geography and literature, taken by the teachers and professors. No more than three or four would gather in a corner for these informal classes which were, of course, forbidden. A look-out kept an eye on the Judas hole and at the first sign of a guard we dispersed immediately. Periodically, the guards would initiate a search of the cell for 'illegal materials' such as letters or chess pieces. During the searches, which were as much about harassment as security, we'd be made to stand naked in the freezing corridor for up to two or three hours. The only things they ever found were those pieces of black dough which served as chess pieces.

The claustrophobic conditions in prison were getting to us all and we were looking forward to a certain degree to getting to Siberia. A journey, any journey, would be a change and at least in Siberia there would be fresh air. We used to say to the guards 'Hurry up and send us! We're fed up in here!'

One guard, who obviously knew the score far better than we did, said 'I wouldn't be in such a hurry to get to Siberia if I was you. Sure, you'll eat and you'll maybe live but I'll tell you youngsters one thing – you won't give a single thought to women when you get there!' He was right, of course. Nor did we realise just how fresh the Siberian air would be!

It may seem naïve, but we had a kind of romantic vision of Siberian exile because Tsarist Siberia had been a place of 'internal exile' rather than of slave labour and many Poles had been sent there. But we hadn't realised just how radically communism had changed, in two decades, what 'Siberia' actually meant. That was the shock awaiting us at the railhead, the camp distribution point.

What was worrying us at this point was only the weather, the permafrost, the minus-50 degrees temperature. That's what we thought we wouldn't survive. Being fed quite well, comparatively, in prison, we had no idea of the meagre rations awaiting us, nor of our conditions of slavery. And we had not the slightest suspicion of the enormity of the camp system – the Gulag.

The camps had, of course, for a long time been part of a grand strategy in Stalin's head. The first camps were set up to build the White Sea Canal. Some 600,000 slaves dug that canal in sub-Arctic temperatures with no earth moving machinery beyond shovels and baskets. Not only did Stalin see this forced labour as essential to the Soviet economy, it was also part of his strategy to prepare for war with the West. The canal was intended for shipping the Murmansk-Archangel fleet into the Baltic. Vorkuta, where I was bound, was different. Vorkuta was about coal.

Realising that war with Germany (and/or the West) was inevitable, Stalin saw his heavy industry, which was largely based in the Donbas in the Ukraine, as vulnerable: quite within range of enemy bombers. Much of this industry was moved east beyond the Urals in the mid 1930s. But you couldn't move the Ukrainian coal deposits upon which this industry had been built. Fortunately for Stalin (and unfortunately for me and my fellow prisoners) large coal deposits were discovered around Vorkuta and our job was going to be the construction of a rail link between Vorkuta and the heavy armament furnaces in Magnitogorsk and Chelyabinsk – east of the Urals and well out of bomber range.

Incidentally, the man in charge of heavy industry at the time was an old pal of Stalin's, Grigory (Sergo') Ordzhonikidze. It was Ordzhonikidze and Stalin who (without Lenin's approval) ordered the Red Army into Georgia in 1921 to bring it under Soviet control. As Commissar for Industry Ordzhonikidze did the job alright, moving the heavy plant eastwards, but he was also concerned that Stalin was bumping off the Old Guard during the thirties, murdering his old comrades. After the second of the great show-trials in January 1937 and an angry confrontation with Stalin, Ordzhonikidze shot himself (or was murdered on Stalin's orders) on February 19th.

Stalin's policy and Ordzhonikidze's organisation proved their worth during the war. Vorkuta coal helped the USSR exceed Germany in the production of tanks and weapons. On our way to Vorkuta in 1941 the strategic plan was obvious to our eyes. Camp after camp after camp to build and maintain the rail link through the devastating winters.

An interesting historical footnote here concerns Soviet-German collaboration and Stalin's other policy – wiping out people close to him. I've already mentioned how vigorously and enthusiastically the USSR supported the Nazi war machine following the Nazi-Soviet pact of 1939. But German-Soviet collaboration went back to the Treaty of Rapallo in 1922, a treaty of mutual assistance.

Both countries opposed the Versailles peace settlement following the First World War and, as outcast nations, they were of use to each other. Through the Rapallo Treaty the Soviets hoped to gain in terms of military technology and strategy while Germany got access to oil and iron ore from the USSR. Germany also got somewhere to develop armaments and carry out military training which, under the Treaty of Versailles, it could not do on German soil. German pilots and tank crews were therefore being trained in Russia. But all this stopped when Hitler came to power because he wanted to end communism.

Was it any co-incidence that Stalin made his moves to purge the Red Army leadership in 1937, by which time he knew that, sooner or later, Hitler would invade? Marshal Tukhachevsky, Deputy Commissar for Defence, was tried and executed for treason, along with other top generals accused of plotting with the German secret service. Historians have argued that this may have just been an excuse for Stalin to wipe out the potential for a military coup against him; that the conspiracy did not exist and/or what secret evidence Stalin had was a Nazi fabrication intended to weaken the Red Army. All of this may be so. But it remains a fact that the older generals had worked closely, had collaborated, with the *Wehrmacht*. What if the Germans had actually bought their services? There remains a possibility, perhaps even a probability, that Tukhachevsky and others were compromised.

Looking at it strategically, Stalin may well have been right. And when it came to the Nazi-Soviet pact, who gained most? Hitler thought he had. He avoided fighting on two fronts but the price he paid was to allow Stalin to move the Soviet border one hundred miles west – a crucial hundred miles when you remember that Hitler was eventually stopped only fifty miles from the gates of Moscow.

Looking at this geo-political situation it seems Hitler was just not in Stalin's league. And what if Stalin hadn't killed off the Old Guard and they had betrayed him? The point remains; the younger officers had no history of collaboration and they proved themselves very able commanders. They won through.

I had a lot of time in prison to think of who had informed on me. With the help of a few trusted cell mates, close friends, I managed to put the puzzle together. Since that day when I'd insulted Stalin by sitting on his portrait, I'd been under surveillance. But who was my shadow? It had to be somebody young, somebody who shared common interests and who could get close to me.

I had a lot of sporting interests, as I've said. When the Soviets occupied Luck they had closed all Polish sports clubs and established two all-embracing clubs: 'Dynamo' and 'Spartak'. The first was exclusively for NKVD employees, the second for Soviet Army personnel. A few locals were admitted to both as 'window-dressing'. One of those admitted to the Dynamo football club was a young, local Ukrainian whom I'd met casually through a mutual friend. I remembered now that he had become unusually friendly with me and some of my friends. I was too young and inexperienced to suspect anything.

He had become my constant companion, my shadow almost, promising to get me into the Dynamo football team. He had taken me several times to the Dynamo river sports club, no doubt to show me off to his handlers, to demonstrate that he was 'working' on me. In order to cover his tracks he'd told me how he hated the Soviet occupation and that he'd joined a secret Ukrainian underground organisation which, at the appropriate time, would fight alongside the Polish underground to destroy Soviet rule. He infiltrated my circle of close friends and took part in our activities and discussions, no doubt reporting regularly to the NKVD.

Then, one day, one of those close friends disappeared. Rumour had it he'd been arrested. But after a few weeks he suddenly reappeared. This should also have rung alarm bells in my head. Only those who co-operated were released. I ignored this as he was such a close friend. I thought that maybe his parents had some influence and got him released.

He rejoined our group and, after a few weeks, when we were alone, I asked him if he'd like to re-join the underground. He refused. But my goose had already been cooked anyway.

Both these informers survived the war and settled in central Poland. What do I feel about them today? If it was my destiny to go to Siberia, then they were only the instruments of destiny. So be it. Both of them are long dead. I am still on this earth.

Around midnight one night, just prior to my deportation, the cell door opened and an NKVD officer called out my name. I was taken to see the prison commandant who was sitting in his office with papers spread out on the desk before him. He had the assumed air of someone considering something deeply. He looked up.

'Within the next few days you will be on a transport to Siberia.'

He paused.

'Only, you don't have to go if you are willing to co-operate.'

Co-operation meant informing, pure and simple. We all knew there were informers in the cells and so no one spoke to a prisoner they didn't know well. A little angel on my shoulder was whispering 'Whatever you do, don't co-operate'.

'You'll see your parents once a week and you'll receive food parcels from them.'

This was a tremendous temptation to an eighteen year old. I had only a few seconds to make up my mind, but my little angel had already worked his magic.

'Thank-you, citizen commandant, but I'd rather go to Siberia with my friends.'

'That's your choice' he said 'but you do know you won't survive there.'

This was a pivotal moment. Ironically, it was one that ensured my survival, as I found out when I first met my mother again after the war. She was never informed officially of my whereabouts and never knew if I'd been sent to Siberia or Kazakhstan or held in a local jail. But when the Germans invaded the USSR, the NKVD shot every prisoner in Luck and the surrounding prisons. My mother spent hours looking for my body.

These prison massacres were cold-blooded cruelty on the part of the NKVD. If I'd remained I would undoubtedly have been among the dead. Fortunately, most of the Polish prisoners were already deported. It was mainly Ukrainian nationalists who were left in the prison. They were executed because the NKVD knew that many of them would probably have co-operated with the Nazis.

11
Journey to Vorkuta

I had never really been away from home when we set out in early February 1941 for the Gulag. It took one month to get to Vorkuta, a month of gruelling hell. We went by train stopping off at shunting stations and prisons on the way to pick up more cattle trucks, more prisoners. At every stop the train got longer, accumulating over thirty wagons: the same kind of wagons the Germans were using to transport the victims of the Shoah to Auschwitz and their other death camps.

These wagons were no surprise to us. We had seen them used to transport Polish officers and their families to Kazakhstan the previous winter. Each had two doors, two tiers of bunks fixed to the walls, a central stove and a bucket for human waste which was emptied every morning by the side of the rail track. Each wagon had its own armed guard stationed on a platform at the back. Well insulated against the cold spring weather, they wore big boots, padded jackets, gloves and warm Russian hats with ear flaps, the forehead adorned with the ubiquitous Red Star.

The doors were, naturally, always locked. There were two small ventilators above the door but you couldn't see anything. It was always pretty dark during the daytime. We huddled by the small stove in shifts to try and keep warm.

The wagons were travelling cells with the same prisoners kept together all the time. Crushed together in the bunks, if one man turned in the night we all had to turn over at the same time. We were always fully clothed in the winter clothes our families had given us. The smell was incredible, but you can get used to anything. It was only when we stopped at prisons along the way, at Kiev, Dniepietrovsk and Kharkov, that you got the chance to clean up in a *banja*, a pretty basic kind of Turkish bath where you chucked water onto hot stones to make steam and then plunged into a cold water pool.

The first leg of the journey was from Luck to Lwow where we sat in jail, gathering more prisoners and waiting for a new train to take us out of Poland and towards Kiev. It was on this second leg that reality, and sorrow, struck hard. At the border between the old Poland and the Ukraine the wagons had to be lifted onto a different set of wheels because of the different Russian gauge. It was a noisy and threatening process: it symbolised exactly what was happening.

It was difficult for us all. We knew we were finally leaving Poland. Leaving our roots. We didn't know if we would ever get back. I cried. I broke down. Most of us did. We cried quite openly.

Thoughts flooded in. What is happening to my parents? To my family? Will they be transported to Kazakhstan? Will I ever see them again? If they are in Kazakhstan, will they survive? Exiles were simply dumped there, with nothing, to fend for themselves. If Jan and Stanislaw are with mum and dad, they might survive. Dad was no handyman on his own. Or had they dragooned my brothers into the army?

The family had been broken apart and it was my fault. I was marooned in my own guilt. What had I brought on their heads? I still have that guilt today. I was responsible for the break-up of my family. Maybe a lot of my friends felt the same, but we never spoke to each other in these terms. Maybe even today we'd find it too difficult. But the enormity of what I'd done hit me. I'd maybe thought I was a bit of a hero. Was I a hero or should I have kept my head down like some of my friends did? Why had I thought it my duty to join the AK?

From childhood it had been drummed into me that to keep Poland independent was the prime objective of my life. But was it? It was then that I began to re-assess such things. I began to see that independence is a relative thing and that a young man's, any man's, first duty is to his immediate family. You have a duty to ensure that you deserved to be brought into this world. *Honour thy father and thy mother that thy days may be long upon the land.*

Sure, my father understood why I was in the AK, why I'd joined. But I think my mother was always terrified I'd be taken, though we never spoke about it. What anguish was I putting her through now? These were the kinds of thoughts I was going to live with for a long, long time. I live with them still.

I need not have worried about dad as he was in a safe occupation. But throughout my exile I kept thinking about my parents and worrying about them. Under the Soviet terror, if a person was arrested their whole family usually suffered deportation and confiscation of all property and possessions. But there were certain exceptions for a limited period of time: people whose skills were badly needed like doctors, dentists, certain engineers and technicians. My father was an expert in the financial and telecommunications system the Polish Post Office had acquired from the West and installed in 1933. He was essential to show them how the system worked and to train Soviet personnel. He also spoke excellent Russian. The integration of the Polish system to the Soviet one was never completed because of the German invasion in June 1941. The Germans also needed my father's skills.

In the prison in Kiev I met a certain General Mirsky, a man without whose fatherly protection I would not have survived the Gulag. Mirsky had led the Red Army in the Urals during the Civil War before becoming Secretary General of the Communist Party in Odessa in the Crimea. He did not know what he had been arrested for. Few of the Russian people I met in the Gulag did. He was in his fifties and of Polish origin. I trusted him. Sometimes you just get a feeling. He spoke openly and didn't seem to hide anything.

'I'm not aware of having been involved in any counter-revolutionary activities, but that's what I'm accused of', he said. 'Somebody must have fingered me. Maybe I said something unguarded. I don't know.' Given his military rank and his party position, it must have been a powerful enemy – or enemies – who denounced him.

He asked me where I was being sent. When I said 'to the Kotlas region' he turned white and said nothing. Kotlas stretched from Kirov to beyond the Arctic Circle where Vorkuta was. I don't know whether Mirsky turned white out of sympathy with me or whether he had just realised where he was going too.

Another prisoner in the cell kept declaring 'It's a mistake. I'm the Minister of Trade for the Ukrainian Republic. I'll be got out.' But as the days went by his face grew ever longer as he began to accept that his 'friends' would not rescue him. This was a typical Gulag experience. Quite simply, if Stalin issued an order that he needed X number of people to construct a railway or the canal, the NKVD had to find the labour. Innocence or guilt were irrelevant.

Each wagon had its spy or collaborator, most of whom were Soviet officers. We had a lieutenant who joined us in Kiev. When he entered our cell we asked him, as we all asked each other, what he was in for. 'I don't know' came the answer. He never gave answers but he asked a lot of questions and listened in to our conversations. He was a Ukrainian, in his thirties, and he understood Polish. It was two or three days into our journey from Kiev that Mirsky said 'Watch him. I don't trust that he's a genuine prisoner.'

This lieutenant was always trying to organise things in the wagon, the distribution of rations, the rota for cleaning the waste bucket and so on. It was as if he couldn't not control, maybe a result of his army training. When we got to the end of the line, the railhead in Siberia, he disappeared and was never seen by any of us again. We knew for certain then that he had been our spy. He may not have been our only one.

In the holding cell in Luck I'd been befriended by a Pole who'd been second in command of the regional police in Luck before the war. He was also on our transport and wore his old police uniform all the way to Siberia. I never thought about it at the time but it was strange that such a high ranking policeman was still alive and hadn't been transported to Kazakhstan earlier with the army officers and other police. He got close to me and I shared my bread with him. He always said how much he owed me. Yet when I corresponded with him when he was in London after the War he seemed to want nothing to do with me. He said he had been on General Sikorski's staff and that he was a fingerprint expert (though I don't see how that is meant to add together). He died some years ago. Thinking about him now I can't help wondering: was he one of Stalin's men in the Polish Government in Exile? I've never been able to figure him out. A strange survivor, indeed.

There were stops, sometimes for days, in various prisons in Lwow, Kiev, Dniepropietrowsk and Kharkov before we travelled on to Siberia through Moscow itself. On station platforms or as we were marched through the streets to and from these jails people stared at us. It was obvious they were aware of *what* we were though maybe not of *who* we were. They stopped and watched in silence. You could see in their eyes that they sympathised but could not say or do anything because the Soviet State was the NKVD State. You could tell they were used to seeing these processions and knew what our ultimate destination was.

In the prison in Dniepropietrowsk, in March 1941, we were actually given newspapers. It was the one and only time and it was typical of the curious anomalies of the system. I read in *Isvestia,* incredibly enough, that the Germans were conducting huge land manoeuvres on occupied Polish territory, close to the new partition border with the USSR. I was filled with hope immediately. We all knew, the Polish prisoners that is, that this could only mean preparations for the drive east. As things couldn't be any worse for Poland or for us we thought 'Come on Hitler! Do your worst!' For once these two bastards, Hitler and Stalin, were at each other's throats there was hope for the cause of Poland, for the Allied cause and for us.

The report was unbelievable in a way; not that we thought it lies, though lies you would expect in papers like *Pravda* or *Isvestia*. But it was obvious they couldn't see the import of what was happening. It was reported along the lines of 'Our brave German allies getting up to their own big manoeuvres'. Were they blind? Couldn't they see? Where did they think these battalions in Poland were headed?

Of course it's now well recorded that Stalin was blind, perhaps wilfully so, to Hitler's treachery (though you might have expected one megalomaniac psychopath to have some understanding of another) and did not even believe his own intelligence reports in the first hours of the actual invasion. Sometimes Soviet stupidity was quite astounding.

We talked politics quite a lot, sometimes just to relieve boredom or monotony, and both Mirsky and the spy joined in. The show trials, the Great Terror as it became known, was naturally a hot topic. I remember discussing the execution of Yagoda who had run the NKVD until 1937 and been shot in 1938. Mirsky's take on why Yagoda had been killed was that he had been planning to arm the inmates of the Gulag, people who had nothing to lose, to challenge for power.

While the journey to Moscow was broken by these prison stops, the monotonous journey north from there onwards seemed interminable. On and on, day after day, night after night. The only time the doors opened and you saw the outside world was at morning slop-out and in the evening when you received your black bread and water. This was when you got a frightening glimpse of where you were – and where you were headed. You began to notice how the terrain was changing, slowly, from forest *taiga* to Arctic *tundra*. And always, at a distance, camps and more camps, huge camps. Never towns. Never villages. Just camps. Only then did the enormity of the beast really begin to sink in. There were millions of people out there – out here – in these camps. Millions!

And if that were the case, there must be thousands of NKVD guards to run them. Perhaps hundreds of thousands. But the Soviet economy couldn't afford that. The Soviet system wouldn't allow for such a waste of potential manpower. It didn't make sense. But, if they didn't have thousands and thousands of NKVD guards, how was the Gulag policed? We were soon to find out.

Vorkuta

12
Walking Shadows

Finally the wagon doors opened for the last time to the bitter cold and the snow. Here, you would never take your clothes off. Ever. It was too cold even in the summer. It snowed all the time.

The whole train load of us were now marched one mile over the frozen tundra from the bleak railhead to the Pieczora dispersal camp. The camp consisted of wooden barracks, very similar to those the world knows from Auschwitz-Birkenau, and you could see that they had been recently constructed. Business was obviously booming in the Gulag. Each barracks held two to three hundred people and there were some Poles already installed in ours. They had arrived a few days before on an earlier transport and they were the first to warn us.

One of the first things we noticed, because it was very strange, not to say astounding to our eyes, was a group of men sitting on a bunk playing cards. Such things just did not happen 'inside' the system. No prisoner could get away with this kind of behaviour.

'Be careful' one of the Poles said, nodding in their direction. 'They are your lords and masters here, not the NKVD. Get on your bunks and don't move. Don't attract their attention. They'll take your boots and your clothes if they want them and if you refuse, they'll chop your head off.'

'Who are they?' I asked.

'Criminals. They run the place for the NKVD. They're cruel and they're encouraged to be as cruel as they want to us.'

By 'us' he meant the political prisoners.

'They have knives and axes and they'll use them.'

We clutched our precious bundles and sat on our hard bunks trying not to move.

Fortunately, the next morning we were detailed to a further transport. We trudged two miles to a river, the River Pieczora which drains into the Arctic Sea, where we boarded barges for the last leg of our journey. A single tug pulled three or four barges at a time, each barge holding some two to three hundred men, all bound for a life, and most probably a death, in the frozen heart of the Gulag. We spent three days on these large uncovered barges with only a little black bread, pickled herrings and water to keep us going. All

All the digging was done by hand

the time we kept a wary eye open for thieves. To lose what clothes you had in your bundle now was as good as to lose your life.

Once or twice a day groups of some forty to a hundred men would be disembarked to disappear towards yet another isolated camp. Watching these men go I began to realise that the chances of my ever returning home were virtually nil. It would take a miracle and I wasn't sure that I believed in miracles.

Eventually, we came to a stop south of Vorkuta. It was our turn. The forty of us, all Poles, disembarked along with General Mirsky. I remember one man looking reasonably warm, an NCO from a Polish tank regiment who was still wearing his army issue leather coat. We tramped for three or four miles before our camp came in sight. A big gate. A watch tower. Barbed wire. A few buildings and shacks. A couple of big field tents.

'Tents?' I thought. 'In these conditions?'

The gate opened. We were in. The final destination.

Roll call. An NKVD officer appeared, alongside him the camp administrator who, it turned out, was a prisoner of German descent. I never knew what his 'crime' was, nor why he was chosen for his position.

'Now that you are here you should know that you have no rights except the right to work', the officer told us.

'Every day you will be going outside the camp to work, to dig, to cut wood, or to carry supplies from the barges to the camp. If you are obedient, if you are a good worker, you may be rewarded with a job in the bakery or the camp stores.

'If you are thinking of escape, forget it now. Look around you. There is nothing out there. You won't survive. You won't even get very far.

'The people who live out there, there are a few of them, are Samojed.'

These were Arctic tribespeople who were paid handsomely with a piece of pig or reindeer or a small bag of flour to return any escapees, dead or alive was irrelevant. Their name roughly translates as 'cannibal'.

Everything, it was grimly apparent, was organised to the *nth* degree to ensure we would remain where we were until we were no longer of use to our masters. We were now taken to a dug-out, a kind of submerged Nissan hut with earth covering the curved tin roof. Taking five or six steps down, we entered the darkness. Peering around we eventually made out the two and three-tier bunks and the central stove. Our living quarters. There was no light. You knew you would be in total darkness inside and outside, twenty four hours a day. Here we would spend our nights, three men squashed into every tier, all moaning and groaning, coughing constantly and scratching because of the infestation of lice.

If the Russians treated us Poles badly, they treated some of their own far worse. We soon learned who the occupants of the two field tents were. Three hundred and sixty Soviet officers arrested after Stalin's disastrous invasion of Finland. They were regarded as 'traitors'. Their 'crime': getting captured by the Finns. I was befriended by a young Red Army Captain who had been wounded and captured. Arrested with the others on his return to the Soviet Union at the cessation of hostilities, he had asked the NKVD what, as a wounded combatant, he was supposed to have done in such circumstances. 'They told me I should have shot myself' he said.

Inside the dug-out we huddled down for the rest of the day. We checked we still had all our worldly possessions wrapped up in the blanket we'd each carried all the way from Luck. These old clothes, we knew, were priceless: a means to trade and survive. And we were lucky: there were no criminals. Though this was to prove a temporary respite.

The next morning we were divided into 'work brigades' of twelve, each with a brigadier. Our brigadier was Mirsky. So began our daily and nightly task of digging the cutting for the rails to be laid. It was impossible to dig through the permafrost which was a metre deep. It had to be blown up, usually by some of the captive Soviet officers who knew about explosives. Then the rest of us would set to digging and clearing.

Apart from NKVD personnel, the external guards were often Soviet soldiers who had committed some 'offence' in the army and been sent to the Gulag for a minimum of three years. This sentence could be reduced if they were seen to deal with the prisoners severely. But the opposite also held true. A soldier showing leniency or friendship towards a prisoner could have his time extended.

These external guards never entered the compound, which was 'policed' by the criminal element. The external guards' food ration was not much better than the rations received by the 'best working' prisoners. Only NKVD officers got good food and plenty of vodka to keep out the freezing cold.

In those first few weeks we were angry, an anger that I think stayed with most of us Poles until our release. We were angry at what had been done to us and our families, angry at the injustice of it all. But the Russian prisoners just seemed to accept their fate. I didn't know if this was evidence of that famed Russian fatalism or if they just knew too well how the system worked but there was a heavy air of resignation among them all.

As we dug we had to load the earth into large single-wheeled carts then dump it well clear of the cutting. It was hard, heavy work. It took strength and stamina and keeping up both was *the* problem. Your food ration depended on reaching a certain norm: so many cubic metres of earth shifted per brigade per day. The brigade had to reach the norm to make sure we all got enough to live on; so the brigadier had to make sure everyone pulled their weight. Above him was the taskmaster in charge of maybe ten work brigades. Ours was a typical Russian: friendly, hospitable and fair. As an eighteen year old I was the youngest man in the camp and to some degree I was shielded by the older prisoners including Mirsky and the task-master.

There were three categories of food. If you were judged to be giving 50 per cent effort or less you got a small piece of bread and clear soup with maybe a herring head in it for stock. If your work rate was judged at 75 per cent you got a bigger piece of bread and maybe a bit of fish flesh. And if you were up to the 100 per cent mark you got half a loaf and richer soup with maybe even some vegetables in it. At the start I was warned by the Russian officers 'Don't even dream about giving 100 per cent because the energy you expend will never be compensated by the food ration you'll get.'

As always in the Gulag, the name of the game was survival and I never gave more than 50 per cent. General Mirsky, however, always put me on 100 per cent rations. This worried me as I didn't want him punished on my behalf.

'Do you think anyone really checks these figures in Moscow?' he said. 'If they did then they'd realise the railway would have reached Vladivostok by now!'

The work went on twenty-four hours a day and you worked in twelve hour shifts. At night, you worked under arc lights. The only compensation was the few seconds you might snatch to watch the Aurora Borealis: rare moments of haunting, Arctic beauty.

Outside the perimeter fence stood a few buildings which housed the NKVD officers, the guards, the food and clothing magazines and the bakery. They were put there to be out of reach of the prisoners. The kitchens, though, were inside the camp itself close to where the administrator and the duty guards were quartered.

After our first week a new transport arrived: our lords and masters, the criminals, had come. They didn't work, or if they did it was in the bakery or food stores, and they were, as by now we knew, the eyes and ears of the NKVD. I had a good warm coat, lined on the inside with fox fur, and a pair of stout boots my mother had given me. It wasn't long before I was approached.

'I like your coat and I like your boots. Maybe we could make an exchange?'

I was young and maybe looked a bit wet around the ears, maybe he was giving me time, but he knew he was making me an offer that I couldn't and wouldn't refuse.

'I'll find a good coat and boots for you and I'll supply you with bread for a month. Think about it.'

I did. I knew the coat and boots would be coming from a corpse, a prisoner worked to death. So I went to Mirsky and asked him what I should do.

'If you don't exchange he'll kill you and no one here will know when it happened or how. There'll be no witnesses and the NKVD will report a death by natural causes. Take the bread and you'll live a month.'

I took the bread and the new coat and boots. They were inferior but at least I got one loaf a day for a month, though it was poor compensation as I watched him parade around in my fur-lined coat and water-tight boots. I ate half my loaf every day and shared the rest with others.

The co-operation between NKVD and criminals continued as long as the latter were delivering information. As soon as this dried up they were finished. They disappeared to make room for a new batch. Maybe they were moved on to another camp, but camps were so far apart I think this is doubtful. When their usefulness ran out they probably met their deaths (though what has never met its death even in post-Soviet Russia is the co-operation between state security and criminality).

We always tried not to talk in front of them. But it was dark in the dug-out and they could hide anywhere. You had to be careful every waking minute. They were essential to the practical policing of the camp and they also symbolised its spirit. The machine was oiled by fear: no one should trust anyone.

By now the taskmaster knew Mirsky was increasing my quota but he turned a blind eye, maybe because I was the youngest in the camp. Between them they saw that I often got one of the better jobs, pumping water into the laundry or the bath-house or keeping the guards' fire going. Out with a work brigade, the guards had their own fires to keep them warm. No prisoner was allowed near them but my job, on these occasions, was to fetch the wood and feed the fire for them which meant at least some warmth for me. On both these jobs you also got better food because you were freed from the quota system. There was no 'norm' for pumping water and no way to measure how big a fire was.

As you worked at the digging you were never allowed to move more than a few feet away. I remember once the guard marking the limit by pointing to some bushes a few feet to the side of where we were digging. 'If you step over there, even one inch, you'll be shot.'

And it happened. Someone accidentally put a foot wrong and a shot rang out. A man was dead. But for many fear of hunger and fear of the weather were worse than the prospect of being shot. Many of those who'd been through several winters here regarded death as a release.

People became walking shadows. Every morning a roll call of shadowy figures with rags for clothing reinforced the idea that this was your life until death. Sometimes people disappeared into 'hospital', a hospital none of us ever saw and from which no one ever returned. Was there actually a hospital? I've no idea. If there was then it was simply somewhere 'out there' where they took people to die quietly with, or without, a roof over their heads

The two big health fears and the two most common ways for the shadows to disappear were scurvy and night blindness. Both were caused by our starvation diet which was practically devoid of vitamins and of salt. Night blindness means quite simply you cannot see in the dark. During a Siberian winter it is dark all the time. If you can't see, you can't work. If you can't work you can't maintain any quota whatsoever. You are not fed. You begin to starve. Your purpose is served and you disappear to the 'hospital'. This was another way the system finished you off.

People died every day but you didn't see death around you constantly. You saw people deteriorate and then disappear. Maybe the authorities didn't want to demoralise us too much and so the walking shadows were taken away. Perhaps they feared we might give up altogether and refuse to work if we were exposed to the reality of so many deaths. Then they would have had to shoot us all, which would have been somewhat counter-productive for the Gulag 'economy'.

We became desensitised to death, squalor and suffering. We knew death could come any time, though the spirit of freedom and hope never quite left us. After three months, and this was the Siberian 'summer', I began to develop open sores on my legs. Scurvy was setting in and I knew my time on earth was now limited. The 'hospital' was beckoning. I was on my way 'out there'.

13
Comrades!

Then the news came. 'The war is coming! Germany has attacked!' We were elated. We even taunted the guards. 'Now you'll see what for! What a shame there are no trees here because the Germans would hang you from every one! But it doesn't matter, they'll shoot you down like pigs anyway!'

The Russian prisoners were subdued. They knew they were in a no-win situation whatever happened. We weren't stupid enough to think the Germans would 'liberate' us, but we badly wanted revenge on the USSR for the miseries it had caused us and our families and friends.

To hell with 'liberation'. It was hate, pure and simple, that was coursing through our veins; hate for the corrupt, corrosive Soviet system which no longer seemed invincible. Let it collapse. To hell with work. What's the point now? Our work rate fell accordingly.

Of course, if the Germans had come to Vorkuta they wouldn't have released any of us, Poles or Russians. 'Slavs are slaves' was the Nazi slogan and they'd still need Vorkuta coal. Yes, they'd have kept the slaves hard at it too. And then some. But I had reached the stage of resignation. I was resigned to death. I was entering the realm of the walking shadows. Then, one morning at roll call, all the Poles were put to one side apart from the work brigades.

'Comrades!' the NKVD officer addressed us. Comrades? We were numbers. We didn't even have names. What on earth were they up to now?

'Comrades! We have to tell you that we are now allies! Your Prime Minister, General Sikorski and Comrade Stalin have agreed that you are to be released to join a Polish army to fight the aggressor! As we are now allies, you can no longer live in the camp. You are to be taken outside to the guards' quarters. You are to remain there until it is time for you to leave.'

We felt an incredible burden being lifted as we glimpsed our first, precious hope of freedom coupled with relief and joy at the prospect of a new and positive adventure. Despite our hatred of the Soviets we would get the chance to fight the Germans, the prime cause of our predicament. And the Soviet word was, for once, made flesh. Meat and fish, to be exact. Food with vitamins. Within three or four days my energy began to trickle back.

We had been lucky, as we were later to realise. Our camp had a radio which brought the news of the German invasion and of Sikorski's demands for our freedom. Some isolated camps lacked even this contact. In them many Poles died before the word reached them or, when the news finally got through, the winter conditions prevented their release.

The Russian officers were happy for us. But we felt sad for them. We could see the envy in their eyes. Mirsky wished me well and said how glad he was. The sadness in my heart was enormous. This man was my substitute father, part of my life and the main reason I had survived. I hoped for his release. I hoped for all of them. What happened to the General and his captive comrades I don't know. But if they were released, such was the unforgiving system, they would probably have been put in punishment battalions. Very few would have survived that.

Our release was unique in the history of the Stalinist Gulag: so many people set free and their civil rights restored at one and the same time. We had survived. We were no longer numbers, no longer slaves. Our identity was returned to us. We were once again people, individuals.

If my experience of the Gulag, my whole sojourn in the USSR, taught me anything, it was that the Soviet Empire was run not by the Communist Party, but by the NKVD. The NKVD had its placemen (and women) and informers in schools, in the Stalinist blocks of flats (which were perfectly designed for watchful eyes in every stair and on every floor), in offices and prisons; in the armed forces; in the ranks of its own organisation and in the Politburo itself. The whole structure of the Soviet state was based on informers. Historically, this was not peculiar to the Soviet system. It began under the Tsarist regime with the Ochrana security organisation and it remains today in the post-Soviet FSB.

Throughout its evolution the nature of the beast has not changed: from Ochrana to Cheka, OGPU, NKVD, MVD, KGB to FSB. Its *modus operandi* has always involved, and continues to involve, working with criminal elements to eliminate opposition and maintaining power through fear. Any sign of dissent was reported and punished by either deportation or death. Dead people do not talk and they cannot conspire. It was only through fear that Stalin could maintain an empire comprising close to ninety nationalities, seventeen 'republics' and seven satellite states. The system was manipulative and cruel to the detail, creating a defeatist society where people thought not of enjoying life but simply of surviving to the next day. It was not only my life that was not worth the price of a bullet but the lives of the majority of people living under Soviet tyranny, and that would include those in the highest echelons of the Communist Party itself.

It will take a long time for post-Soviet Russia to shed the culture of criminality and fear and it will not do so as long as the present security apparatus exists. Moscow's former Soviet colonies in east and central Europe are only now beginning to break through these barriers, so deeply embedded in their land. The Russian power system remains primitive and cruel and has no rightful place in a modern, democratic Europe.

14
Out of Captivity

Two or three days after our official re-instatement as human beings we marched back to the Pieczora Lagera to board the open barges again, this time without the kindly attentions of any NKVD guards. We had been given official documents stating who we were; that we had been released under the amnesty to all Polish political prisoners and that sentence passed on us was thereby annulled. These were signed by the camp commandant and the NKVD officer in charge: typical of the Soviet system where everything was signed twice. In any case, without the NKVD signature the documents wouldn't have been worth the paper they were written on.

The barges were filled only with Poles and, despite being wet, cold and hungry, we were in high spirits when we arrived back at the railhead barracks, which were now also blessedly free of criminal attendants. Here, I was issued with a cotton-insulated jacket and new boots and we all received ten or twenty roubles and half a dozen pickled herrings wrapped in newspaper. We were hungry as hell, as you can imagine and I, rather stupidly, ate all my herrings in one day.

We were then loaded onto the cattle wagons, but this time they remained unlocked and there were no guards. As we began to roll slowly away we burst into song: Polish patriotic songs, to be exact. Then we remembered that such songs were forbidden. So we sang them all the louder and all the longer.

For the first three or four days, after my herring feast, I was absolutely desperate for water. At one stop where the train was taking on wood or coal I drank from a stream. I didn't know it was polluted but even had I known I think I would have drank anyway. The result was stomach trouble that lasted for a year; not dysentery but plenty of vomiting and diarrhoea, which was common among many of us in any case.

I travelled in the same wagon with Kazek Sarnowski, Kazek Mankowski and Zbyszek Kretowicz, friends from Luck who'd been arrested at the same time as me but who had been in different camps. Our journey to Uzbekistan was to last one month. All along the Urals to Yekaterinburg (where Lenin had Tsar Nicholas and his family executed), on to Magnitogorsk, Chelyabinsk and Oranienburg and thence into the emptiness of Kazakhstan. Those roubles and herrings were all we were ever issued with for that journey and they ran out soon enough. We couldn't buy food. Whenever the train stopped we'd go scavenging in open country or nearby fields and on some lucky occasions we'd dig into winter pits of potatoes and roast them on an open fire.

As our train stopped in Yekaterinburg station another train, heading in the opposite direction, also pulled in. It was a Red Army train full of soldiers in uniform. There among them I suddenly spied my old Jewish friend, Marek Feldman, from Luck. Marek symbolises a strange pattern in my life: of people disappearing and then re-appearing, sometimes to disappear again and, as in Marek's case, re-appearing once more. It was a strange conjunction, anyway, meeting an old school friend a thousand miles from home in a Red Army uniform as I was on my way out of the Gulag! We greeted each other heartily and I asked him what he was doing

'We're running away from the Germans'.

I said for him to join us: 'Join the new Polish army!'

He was keen. He agreed immediately but said he had something urgent he had to attend to first. He disappeared along the line. I waited but he never appeared before we pulled out. I don't suppose I ever expected to see him again. I presumed he would have perished in that terrible campaign: 'the Russian Front'. But I met him again by chance in Warsaw in 1998. He had become the Vice-President of Polish Jewry in our homeland. I asked him what had been so urgent in Yekaterinburg. He said 'I was searching for my girlfriend.'

'Had you come with us', I said 'you would have been in Palestine and maybe met up with Menachim Begin and Moshe Dayan' (both of whom were in our Polish Army).

Marek never found his girlfriend. He returned to Luck after the war before eventually settling in Warsaw.

Once, just over the Kazak border, we saw a glorious sight: a mountain of watermelons. We ate our way through the lot. Being mostly water and very little sugar, they didn't do our stomachs much good. I've never been able to eat watermelon to this day though, I have to say, I still relish pickled herrings (perhaps because I've eaten them since childhood).

It was a difficult journey of slow and slower, of stop, start and stop again. Priority was naturally being given to military and supply trains crossing our southerly route and heading westwards to the German front; and the Soviet rail network wasn't exactly extensive.

We went through Khiva and Alma Ata, where Trotsky had once been exiled. I can still clearly picture the blue mosques of Samarkand. When we arrived in the famous Islamic centre of Bukhara we made our way to the market place where, naturally enough, the free market still flourished. I still had the new boots and jacket issued to me at the Siberian railhead and, as we had no money to buy food, I foolishly allowed my friends to persuade me to trade them.

'We'll get brand new uniforms soon' they said. 'Just think of the good dinners we'll have meantime!'

Well, we did have several good dinners but we didn't get our uniforms as soon as we expected. When we pulled into a siding at a place called Buzuluk where part of the 7th Division Polish Infantry was being formed under General Okulicki we were told there were just too many new arrivals to be processed. So we were to be dispersed to Uzbek villages in the meantime. Not good news. It meant another freezing barge journey – those bloody barges again! Even so far south the icy blasts from Siberia and the steppes of Kazakhstan made the evenings very cold, especially if you didn't have a warm jacket or good boots. It was a two day journey with one stop during which we got some hot soup that revived me a little.

At our destination we were met with several large-wheeled carts each to carry a dozen of us to a separate village. At our assigned village we were met by a party official.

'Dear Allies! (we were getting used to this address by now). Our brave young men are away fighting the common enemy, so you must help us in the fields.'

And what fields they were! Central planning in Moscow had turned Uzbekistan into a giant cotton plantation irrigated by two rivers which were draining the Aral Sea, to the extent that its shores have now retreated around fifty kilometres, destroying the local fishing industry. This policy began around 1933. Everywhere you could see irrigation canals drawing from the rivers. It was now the cotton gathering season.

'The more cotton you gather the more food we will give you' said the CP official.

Having slaved in the Gulags we were wise enough to the Soviet system by now. So we asked 'What's the norm?' and were told 'One hundred kilos per day per man'.

One hundred kilos of coal wouldn't take long to bag; but one hundred kilos of cotton as light as fluff?

'Listen!' we said, the ungrateful wretches that we were, 'We came to join the army, not to pick your bloody cotton!'

The result, of course, was that our rations were low in quality and in quantity. We got flour and oil, and that was it. The oil was unrefined, extracted from some kind of coarse grain, and it didn't do my stomach much good. But that was all they seemed to produce in the area except for grapes which went to party officials and Soviet army officers. Life here was primitive. Most of the locals lived in round mud huts with a central fire which smoked through a hole in the roof, though when you looked into the huts the carpets were indescribably beautiful. The only real building was the school house in which we were billeted. It too had only one room but with an oven in the corner where we could bake our rough bread which, smothered in unrefined oil, was what we lived off for a month.

I don't think living conditions here had changed since the Middle Ages. Only officials and local headmen seemed to have anything approaching a 'decent' life. There were one or two cows to be seen. The only benefit we got from them were dreams: dreams about eating steak and drinking fresh milk.

We were desperate to get into the army. But on our return upriver we weren't allowed into the camp for a second time, because of an outbreak of typhoid. Instead, we were set to work at the railhead unloading sacks of rice and grain. I remember taking an eighty kilo bag of rice on my back when my knees just buckled and I collapsed. I had absolutely no strength. From then on I was left to lift only smaller bags.

We got some rice and mutton to eat and within about ten days my strength began to return. After two weeks, when quarantine was up, the camp gates were opened to us and we were given a medical. This wasn't to see if we were fit for service, which most of us weren't, but to identify who needed hospital treatment. Many of us tumbled straight into the hospital with jaundice, night-blindness, scurvy and malaria (the last being endemic in the area).

We were also issued with British Army fatigues, a symbol that the USSR had washed its hands of us, though we were still thinking we'd have to fight alongside the Red Army on the Eastern front. We still didn't know what was really happening, nor what our ultimate destination would be. Then some of our number were loaded onto trains for Persia (Iran) which was at that time divided into a Russian and a British zone. We suspected they might be heading for the British zone, for the free world, and we were jealous.

What was actually happening was that the Polish forces in the Middle East were being upgraded in strength from brigade to division status, from around 2,000 to around 10,000 strong. The Carpathian Brigade, who in 1939 had made it to Syria via Romania, the Balkans and Istanbul, had gone on to fight at Tobruk as part of the Australian Division. At the fall of Tobruk, around the time we were being released from the Gulags, they'd been moved to Egypt and Palestine. Our numbers were going to help create the new division.

Meanwhile, given my general health, I was assigned to a supply company as a clerk. The whole camp was full of British tents, lorries and equipment, alongside a few Russian rifles and revolvers with, of course, no ammunition. Stalin's fear of his own troops, his own people, was such that Red Army units didn't receive their live ammunition until they were at the Front. And they were given it by the NKVD whose units were always just behind the Front in case anyone should think of deserting.

There were no British officers in the camp but there were Polish officers who'd come over from Britain (most of those who'd been in Soviet territory in 1939 had been murdered at Katyn).

I was beginning to return to health mentally and physically but I was still troubled, as we all were, by how we would fight alongside the Soviets. After all, many of us were sworn into the AK and sworn to oppose Soviet rule. The depressing reality was that we knew it was not only the quickest way home but probably the only way. We were sure of Allied victory eventually but we didn't want the Red Army in Poland and we certainly didn't want to be responsible for helping them on their way there.

I'd been clerking away for around six weeks when my CO told me 'There's an officers' training school being set up. I think you should apply.'

So, I did, even though I was only twenty. Camp life had been pretty relaxed but at the OTS this was to change. Here the discipline was harsh and demanding, as it had to be. We had to learn tactics. We had to learn leadership. We also learned how to trade with the local Uzbeki communal farms. They had grapes and pomegranates and we had soap, so we bartered to improve our diets while bringing some cleanliness to their Socialist paradise.

The OTS was housed in small tents between some of which ran the narrow irrigation canals for the cotton fields. One of our tasks on night duty was to fish men out of them, men who were still suffering from night-blindness and who fell into these criss-crossing ditches on their way to and from the latrines. This night traffic was pretty heavy: many were suffering from dysentery, jaundice and/or malaria. So these night accidents were regular.

As the camp was swarming with mosquitoes, I contracted malaria myself but there was quinine available, thanks to Britain, and the condition was kept reasonably under control.

It was autumn 1942 and the battle for Stalingrad had begun. We were resigned to the probability of being shipped off there, when things suddenly changed. We were told we were going to Persia!

Imagine the relief and the excitement. We were headed out of a state based on fear and punishment towards a normal civilisation, albeit one still at war. On the other hand, we had no idea what our evacuation would actually mean for our future, nor where we would eventually end up. The world was at war and we were without a country of our own.

First we were taken by rail to Krasnovodzk on the eastern shore of the Caspian Sea opposite Baku. I will never forget the sight that greeted us when the train pulled in at the quayside. Hundreds of Soviet soldiers, many with arms or legs missing, many blinded, all crying out for pity's sake, crying for help. This was the cruelty, the human devastation we've come to know as the Eastern or Russian Front. How we thanked God we weren't on our way there.

Next to ours was a train full of German POWs. I thought 'They're going to take our place. The meat grinder must still be fed.' And I thought 'Few of you will survive.'

We felt pity for both, for the wounded Red Army soldiers who'd have a tremendous struggle to get bread in a land without peace and for the Germans because we knew what was waiting for them. Here you could see the desperate plight of ordinary people being driven by the two great maniacs battling on the Eastern Front.

I also saw another poignant sight. It was Major Berling standing at the quayside saluting every Polish troop ship as it left harbour. I didn't realise then that he wasn't joining us. Nor why.

Arrested and imprisoned by the NKVD, Berling had been one of that group of officers who had signed the memorandum to the effect that Poland and the USSR should be allies. As a result he and his fellow officers had been spared Katyn. When the battle for Stalingrad was over, Stalin was to allow the formation of a Polish army, which became known as the 1st Polish Army, under the control of Soviet officers but with Berling, now promoted to General, in charge. As he stood at the quayside Berling must have known that when the truth about the memorandum came out he would not be accepted in the west, not by Sikorski, Anders or any of the Free Polish Forces. His salute was his farewell.

Berling was a man with a deep inner conflict, and things were not to get easier for him. When the Warsaw Rising started in 1944 Berling's troops were at Praga on the Vistula, opposite Warsaw itself. Without consulting his Soviet superiors he ordered them to engage the Germans to aid the Rising. The order was countermanded and Berling was dismissed. The Soviets didn't want the Warsaw Rising to succeed.

Once on board our ship I was cheered to meet up with yet more friends from Luck, friends I hadn't seen since our deportation, including Freddy Sodowski who was to become very important to me later in my life. We crossed the Caspian, from east to south, to the port of Pahlevi, still in the Soviet sector. However, we were received there by British officers, which was reassuring. They told us to shed and burn our uniforms. Then we were deloused. We needed it!

We were issued with clean uniforms in the transit camp where we were also given better rations including bully beef, conserves (unbelievable!) and dates, dates and more dates. The novelty of the dates, the sweetness, the goodness. I gorged myself – and I've never been able to eat dates since.

My malaria flared up again. While my colleagues were loaded onto lorries I had to take my place on the 'blood wagon', the military ambulance, to be transported to Tehran in the British sector. But we were halted at the demarcation line by Soviet troops who searched all the transport. What did they think was worth our stealing? The search was ridiculous. But it was the Soviet way. Trust no one. Search everything. We held our breath and prayed that nothing would happen, that no one would do anything stupid, say anything out of turn. The engines re-started. The convoy moved off. We were leaving Soviet Paradise.

15
Towards the Promised Land

Crossing the border we met a huge convoy, a seemingly unending stream of American Studebaker lorries churning up an ocean of dust. Each was marked with the large letters 'UNRA' (United Nations Relief and Rehabilitation Administration, formed in 1943 in advance of the UN itself). They were loaded with grain and food supplies, weapons and ammunition, all bound for the Soviet Union. While they had a short run to the border we had just crossed, we knew that our journey back to Poland would now be longer. We were still hopeful that after Germany had been crushed the Western Allies might force their way further east, allowing us to fight our way back into Poland.

I knew how lucky I was to be crossing this demarcation line between death and life. I thought back to that fateful Christmas Eve in Luck and to my sentence. Having come through the Gulag system, and seen something of how the Soviet state operated, I now realised that the length of my sentence had been irrelevant. It didn't matter whether you got five or ten years - every sentence was meant to end in death. 'The dead don't speak' as Stalin was fond of saying. For 'enemies of the people' death was always the intention.

Fortunately, I hadn't realised this at the time. If I had I think my will to survive would have diminished greatly and I would have surrendered to the Arctic Gulag. Then I thought of those poor Soviet soldiers in the Gulag, the ones who'd fought the Finland campaign, and I remembered their fatalism. They *knew*, I realised. They had always known the meaning of their sentence.

We were among the last transports to leave the USSR bound for the Middle East in October 1943.Of the 1.5 million Poles deported to the USSR only about 200,000 of us were being evacuated out at this time, partly because winter was fast approaching, making transport from Siberia southwards very difficult; and partly because the critical situation at Stalingrad meant all spare transport was being commandeered for that Titanic struggle.

On arrival in Tehran I was taken to a military hospital where I was laid up and fed on quinine for two weeks while my unit went on without me. When I caught up with them we were billeted in a factory that had been built by the Germans around 1937 or '38. It was an armaments plant with unused machinery which seemed almost brand new. Undoubtedly built to produce arms in contravention of the Versailles Treaty and as part of the putative Berlin-Bagdad axis it had never actually got to work.

After a few weeks in the factory we were taken by lorry into Iraq, then a British protectorate. Here, the Polish Division, which had moved down from Egypt and Palestine, was guarding oil refineries around Mosul, Kirkuk and Khanaquin. We set up the OTS outside Khanaquin and were now busy being trained in desert warfare. Here we met the most enterprising Arab traders and the most wonderful Arab thieves.

Every time we went into the desert on exercise the traders would pop up from nowhere with oranges, Muntaz cigarettes and containers of *arak,* the local hooch. The Muntaz were good smokes. The *arak* wasn't too pleasant, though it tasted of aniseed, but it wasn't too dangerous either. (It was also the colour of weak tea and sometimes the traders would cunningly substitute cold tea if you weren't watchful).

If trading was their daytime occupation, thieving was their nocturnal hobby. Every night they would come and steal weapons from our camp. We doubled the guards. Still they came. Still they got through. We tripled the guards. But to no avail. The weapons kept walking and with them a substantial amount of ammunition. Until one day our CO said 'Stop this music!'

We dropped the doubling and the tripling of the guards. It was making no difference anyway. Instead, once every few weeks, we simply raided the local villages and took our weapons back. The whole thing was pretty good humoured and I don't know of anyone killed or wounded in this bizarre ritual of theft, recovery and theft again. The locals were simply fascinated by guns. They loved firing them off. No wedding or other celebration was complete without the rat-a-tat-tat of gunfire. I don't think they were ever used in anger (or crime) and there was certainly no political motivation.

It was in and around Khanaqin that I saw the first fighting troops of the western allies, including the Black Watch. They were already experienced in this war and were also fighting troops in every sense of the word. Khanaqin was generally out of bounds, partly because whenever troops went on the spree there they usually ended up fighting: the Scots with the English, the Brits with the Americans. It was a relief from the monotony of desert existence and, of course, an inevitable result of the demon drink. The fights were always broken up by the Military Police and the culprits banged up in the 'cooler' for a few days.

On a more serious note, it was beginning to occur to us why it was in British interests to have what were now three Polish divisions in Iraq. We were being held in reserve to take on Rommel if he broke through at El-Alamein and to prevent him capturing the oil-fields and completing the Berlin-Bagdad axis.

I was in and out of hospital every few weeks with recurring bouts of malaria and it was getting to me. I asked the doctor if there was a cure.

'No', he said. 'It's a parasite in your blood. Only a change of climate will help.'

Fat chance of that, I thought.

My stomach problems were also still with me. Until, that is, Christmas Eve 1942, when I went for a swim in the River Tigris and I was gripped by a terrible pain. I managed to reach the bank and clamber ashore. I lay there for some time until the pain eased. And that was it. It never came back. I don't know what happened, though I suspect it was an ulcer that burst. Whatever it was I hadn't simply swum the Tigris, I had swum the River Jordan to the Promised Land of Better Health!

Having completed my officer training I was now waiting to be assigned to an infantry unit when, as it happened, I saw a notice in HQ: 'Volunteers Required for Parachute Brigade'. I asked where the training would take place and I was told: Britain.

'Britain?' I thought. 'Change of climate', I thought. 'That's for me. But how to get there? Who cares. It's nearer home and a better climate could only help alleviate my symptoms.' I passed the medical and found myself in the holding area for the journey. I didn't know what I was letting myself in for, but I knew that, come hell or high water, I was going back to Europe where I would feel more at home.

Travelling through Bagdad was the first time I'd seen women in any number since leaving Luck. They may have been veiled but even that was exciting, it was so long since I'd had any contact with the opposite sex.

We travelled through the Syrian desert to Haifa in what was then the British Protectorate of Palestine. This journey was causing a whole mixture of emotions. Here I was coming from the hell of the Gulag and the purgatory of Uzbekistan and passing through the most historical of places like Nazareth. There was no stopping there, though.

In Haifa, where we were camped in a grapefruit grove (the first time I'd ever seen these fruits on a tree) I met more old friends, including Jurek Koc and Freddy Sodowski. I was particularly pleased to see Freddy yet again, as he had really taken care of me in the prison in Luck after my sentencing.

We were two or three days in Haifa but did little exploring as we were intent on catching up with each other's experiences. Then we journeyed through Bethlehem and across the Sinai Desert into the port of Suez at the south end of the great canal. The first sight of the ships passing through was quite amazing to me.

Suez itself was a mix of slums and pleasant villas, the latter often owned by Greek merchants, many of whose families had been there since long before the Arabs. It had the smell of Europe about it and was very cosmopolitan with a lot of French and English people. It would be an exciting place to explore in peacetime, but we had to board ship within a couple of days.

16
A World Cruise

From a distance the *Isle de France* looked quite small but when the barge pulled up alongside her she was enormous: a 44,000 ton, former French luxury liner now kitted out to carry 10,000 troops. El Alamein had been won and some of the 8th Army were going aboard to be readied for the Normandy invasion alongside the Poles and some Indian troops.

It was an internationally run ship, a floating League of Nations. The Brits were in charge but the anti-aircraft gunners were American and the engineers were, naturally, French. As I had no English but spoke French quite well from my schooldays I often talked with the French crew. They were mostly communist and they refused point blank to believe our experiences in the Gulags. We were making it up, they said. They wouldn't accept anything which challenged their idealistic view of Soviet Russia, an attitude I was to meet again among the mining communities of Fife.

We were put down in D Deck, well below the waterline, two or three hundred of us swinging in our hammocks in what had probably been a canteen for the civilian crew. We were inspected every morning and twice a day there was an evacuation drill which taught me quickly enough that in the event of a real emergency my chances of making it up to the promenade deck before drowning were pretty remote. As the promenade deck was empty a group of us got permission to sleep there. I slept better after that.

The Mediterranean was infested with enemy vessels. So, although we knew were bound for the Cape of Good Hope, instead of going between Madagascar and the East African coast the ship headed east of Madagascar to fool the Germans into thinking we were on our way to Australia or the far East. Then we turned west and docked in Durban, South Africa.

Here the welcome was incredible. A lady dressed in white with a red hat sang patriotic songs at the quayside as we all hung over the railings or jammed our heads through the portholes to get a glimpse of her and drink in the atmosphere.

We were two weeks in Durban and allowed ashore during daylight. This was my first real taste of western culture. Durban was affluent, beautiful and modern. On one trip ashore I found myself quite taken with the Natural History Museum. I was particularly fascinated by one exhibit: a huge shark whose open jaws seemed to be a warning of the voyage ahead.

I visited the cinema, the first time since my days in pre-war Poland, and I was struck by the way we were served juice or tea before the film started. Naturally, the servers were black. Separatism or racism was apparent everywhere, much more so than in Persia or the Middle East. We were amazed at the sight of the rickshaws being pulled by Zulus in bright feather headdresses and by how gracefully they ran.

The war seemed very remote. We felt as though we were on holiday. We were also perked up when we heard that 800 South African nurses would be joining the ship's company – though we knew their quarters would definitely be well out of bounds! The thought of those nurses made our pulses quicken but it was also a sign that the invasion of Europe was being prepared for - and that it would be a bloody affair.

Leaving Durban we headed east again, then south, so far south that snow was lying on the decks. I don't think we could have been far from South Georgia before we altered course. We had no idea where we were, and we certainly hoped the Germans didn't, until we anchored off Rio de Janeiro. This was proving to be a world cruise. As Brazil was still neutral we weren't allowed ashore. The city would have had a nest of German spies in any case but it was still amazing to be looking out over Copacabana Beach and the Sugar Loaf Mountain with the famous statue of Christ atop it and to see the whole city lit up at night.

We were there for forty-eight hours taking on supplies before we weighed anchor again. Now, we asked ourselves, are we going to make an Atlantic dash for Britain? Not at all. We headed east once more and our next port of call would prove to be Freetown in (British) Sierra Leone on the west African coast. All this time we had had no escort whatsoever, simply relying on the speed of the ship for safety; fortunately, a top speed of thirty knots was pretty fast.

Our main concern was to relieve our boredom. Every day 'Bingo' was organised for the troops with thousands taking part on the promenade deck. You paid a shilling each to join in so the prizes were quite big. I didn't play. I learned to play bridge instead, a game I still enjoy today.

Naturally, there were chess games on the go all the time. There were also illegal gambling schools. A favourite was a game called 'Crown and Anchor', a board game with nine squares, each with a different image on it including one with a crown and one with an anchor. The 'board' was actually made of cloth so that it could be whisked up and hidden quickly when the Red Caps were on the prowl. It was a very addictive game in which you could easily lose a week's wages. Fisticuffs were not uncommon.

From Freetown we headed north towards the Canary Islands where we picked up an escort of Sunderland Flying Boats which stayed with us during daylight hours. Crossing the Bay of Biscay at night, and unknown to us at the time, all the South African nurses were being woken up and told to dress in case of attack.

Despite the boredom the possibility of an attack was always on our minds. We knew of smaller ships which had gone before us and which had been torpedoed. That our ship was bristling with armaments gave us confidence with regard to surface or air attack, but we had no defence against U-boat torpedoes. The ship was also so high that jumping from an upper deck would have been suicidal. In the event of an attack we'd have had to go down slowly by rope ladders.

The journey from Sierra Leone to Scotland took only seven days. As we passed through St George's Channel I saw nothing, not even the Isle of Man, as I'd been hospitalised again on leaving Freetown with another bout of malaria. At least I had a good bed instead of a hammock but I was sweating then freezing and shaking all the time.

My first sight of Britain, of Scotland, was of Gourock and the hills around it and of the mountains north of the Clyde Estuary. I can't tell you how it felt except to say (and I know Gourock is not the bonniest of Scottish towns) that I knew then how Moses felt when he first looked on the Promised Land. It was the first time I'd seen such greenery since I'd left Poland for the Arctic wastes in 1941. It made such a powerful impression on me that I've carried that image of greenery in my mind ever since – and I'll never hear a bad word said about Gourock as a result!

Our first task before disembarking was a bizarre one. We had to jettison thousands of South African cigarettes we'd bought so cheaply in Durban. They were good quality Springbok cigarettes but we weren't allowed to take ashore more than two hundred each. 'Rules is rules', as I was to learn, in Britain; though this one seemed damnably silly in the middle of a war. But there it was. And there they were: tens of thousands of cigarettes bobbing up and down in the waters of the Clyde as some 8,000 troops slowly disembarked.

17
Pittenweem and Parachutes

My first normal passenger train journey since I'd left Poland took me across central Scotland to the little village of Auchtertool just outside Kirkcaldy in the Kingdom of Fife (as I was soon to learn to call it). I was only a few hours in the makeshift camp there, really just tents in a field, before I was taken to the Polish military hospital that had been established at Dupplin Castle (kindly vacated for us by Lord Forteviot) in Perthshire.

Here, instead of quinine tablets, I was given liquid quinine which was much more effective though it made everything I ate or drank taste only of a sharp bitterness. But that was a small price to pay. I was never again to have a severe attack of malaria.

I was three weeks in this hospital, set in beautiful grounds, surrounded by Polish doctors and nurses. For the first time I truly felt I was in a civilised country again and the care and treatment I received began the mental as well as the physical healing process.

1943. Parachute Brigade, Cupar, Fife. Before the Battle of Arnhem

Discharged from hospital, I was given a rail pass and told to report to the 3rd Battalion Parachute Brigade, then stationed at Falkland and the nearby village of Freuchie. The Polish Parachute Brigade, with its HQ in Leven, was commanded by the legendary Major General Sosabowski, who had fought the Bolsheviks in 1919-20 as well as the Germans in 1939. After a couple of days there I was assigned for the 'Pittenweem Experience'.

This was to last for a month, during which the ladies of Pittenweem taught us all about Scottish culture, its food and customs, songs and dances. It would provide a grounding for my subsequent post-war career in the UK. They invited us into their homes and they were fascinated by our story, by how we'd come to Scotland via the Gulags, Persia, the Middle East and a round-the-world trip.

This contrasted with the reaction of the local miners who were strongly communist and, like the French sailors, refused to believe our story. We were inventing it, making it all up – it couldn't possibly be true!

I think one reason the women were more sympathetic was because we were young and many of them had boys in the forces. To some degree we were like substitute sons to them. We were special immigrants who were treated specially. They maybe gave us that special welcome because they felt they were doing it for their sons or hoping that, in a similar situation, their own sons would receive the same kind of hospitality.

In Falkland we went to mass every Sunday in the beautiful chapel of Falkland Palace, which still boasts a copy of the Vilnius icon of Our Lady of Ostrabrama made by Polish parachutists. We also began the normal army routine of square bashing and footslogging. Then we began the gradual, staged process of parachute training to get us battle-ready.

Physical fitness came first with a lot of exercising, marching and cross-country running. The firing ranges were always up hill and you had to climb to them every day. The food was nourishing and we returned to being healthy young men quite quickly.

Then we had to train to get used to heights and to learn to fall without hurting ourselves. This was done at the 'Monkey Grove' in Upper Largo where we trained alongside paratroopers from France, Holland and Norway. At the top of a fifty foot tower was a parachute canopy with a harness attached. Strapped in, you simply jumped from the edge and dropped at the speed you would from a plane. As you hit the ground you had to roll to cushion the fall.

This was the first real test. Discovering a fear of heights, some people were too frightened to jump and they were released to other units. There was no stigma attached to this. You had to try it to know if you could do it.

There was all sorts of apparatus to teach you how to cushion your fall because on a real drop there was no guarantee of how you would actually land. There was a swing with a quick-release mechanism which might drop you forwards or backwards, you never knew which, and you had to learn to protect yourself from injury. There were two weeks of this intensive training to condition your reflexes and make you as flexible, as supple, as possible. It seemed as if every muscle, bone and nerve was being conditioned and there were cases of whiplash, concussion and a few broken legs. Though I was still recovering from malaria I was determined to get through.

Life in Freuchie was quiet except for occasional dances in the Lomond Hall. If you wanted night life (such as it was) you had to travel to Perth at the weekends. As the buses stopped early this usually involved a sixteen mile hike back to camp, which was, I suppose, good fitness training too!

It was at New Year 1943-44 that I was introduced to the Scottish Hogmanay and the custom of 'first footing', which amazed me. A whole lot us had left the barracks to see what was what and I think we must have been invited into nearly every house in the village for a dram. I got back to camp at about four in the morning rather the worse for wear. I did take to the whisky, though it's never replaced vodka as my preferred tipple. Our favourite was Teachers as it had a slightly sweet taste which made it more palatable to us than sharper varieties like Bells.

Early in January my battalion was relocated to Cupar in the centre of the county, the capital of the 'Kingdom'. The Polish HQ was in Ratcluan House. The move to Cupar meant the battalion came together after its split residency in Freuchie and Falkland. There were a couple of cinemas and a good NAAFI facility where the volunteer women served local produce and we could play snooker and table tennis. There were also good fish and chip restaurants where we could supplement our sometimes monotonous army diet. Haddock was, for us, a culinary discovery as the only fish we got in Poland was herring, usually pickled as an *hors d'oeuvre*, or pike and carp which were rather bland and usually heavily spiced.

Cupar was well served by rail and bus, so we could get to Edinburgh, St Andrew's and Dunfermline to see the sights and to socialise. Apart from the intense training, cards, girls and dances were our main pre-occupations. There was also a dance hall above the town library and as there were a lot of young people in this county town, dances were regular. It was at one of these dances early in 1944 that I met my future wife, Janet.

We saw each other at a few of these dances (she was a very good dancer!) and we began to date before 'going steady' as the saying was. Janet was divorced and had a little girl, Janice, whom I soon grew very fond of. I always formed friendships with strong women and Janet was to play a major role in my adapting to life in Scotland. She was very fond of Scottish history and literature and I was learning a lot from her. Even though I wasn't thinking of staying in Scotland at this time – in our minds we Poles were still intent on liberating our country – the ground was actually being prepared for me. Janet had a large family which could help me integrate. So I already had a foot in the local community.

Soon the Battalion was on its way to Ringway (now Manchester Airport) where we were given proper parachutes for the first time and had to jump from barrage balloons anchored at around 300 feet from the ground. You were hauled up in a basket with a hole in the middle through which you had to launch yourself. It was now that you had to learn to control the parachute, to minimise the swings. From the balloons we were then transferred to aeroplanes, Whitley Bombers.

The first jump was scary. There were twenty of us, ten to either side of a central hole through which we had to launch ourselves in turn. The first jump was from around 800 feet, normal battle height, and the speed at which you hit the ground was equivalent to jumping off a lorry going at 30mph. Quite a shock.

Looking into the void as you jump is frightening, but when the parachute opens there's a feeling not just of relief, but of elation – and you want to do it again immediately! You take the elation and forget the fear. It's not a fear of heights as such, not if you've got this far; it's a fear the parachute won't open. But once you've tasted the elation, which is almost orgasmic, you want it again.

To get your wings you had to do eight jumps like this. There were casualties. But to give you a feeling of security you could actually see in through a huge window to where the WRAF girls were folding the parachutes in absolute silence and with total concentration. Your life depended on these girls and they knew it. I think it was brilliant RAF psychology to allow us to watch them at work as we passed by that window.

Unlike the American system, the British method didn't allow for a second parachute. We had only automatic parachutes which were triggered to open three seconds after you jumped whereas the Americans also had a hand-activated parachute in case the automatic one failed. Your life depended on hearing that miraculous 'pop' as your canopy opened.

How you jumped from the plane was also crucial. If your limbs were at all higgledy-piggledy you could easily get caught up in the harness ropes and land very awkwardly, even on your head, and be killed or seriously injured.

Getting your wings meant three shillings a day more pay, a huge increase equivalent to about fourteen fish suppers! But to get your three bob more you also had to be able to run three miles fully equipped in a certain time, while leaping across a number of ditches of required length, then finish your run by throwing a hand grenade a certain distance.

Now it was time to get ready for actual battle. In early June 1944 we were transferred by passenger train to bases in the Peterborough area, north of London. Here, the emphasis was on team work. A company of one hundred would learn to jump from a DC3 Dakota, this time from a side-door. This could be extremely dangerous. The planes had to fly in a very tight V formation because only the lead plane – at the point of the V – had a master navigator. The other planes had to follow; so that when the troops jumped they would land reasonably close together.

In my battalion we lost a complete platoon of forty men when two Dakotas got too close and one went down. This was a terrible waste of human lives. It was also a serious setback for us militarily. These men had been highly trained in the use of heavy mortars: without them, because time was tight, we would have to fly into battle with less well trained substitutes.

During simulated attacks the whole battalion of between four and five hundred men would jump together to land at an airfield with all the personnel watching us, admiring the sight. That certainly was the good side.

Night drops were also very dangerous as you couldn't see one another easily. We were dropped from 300 feet, the minimum height for a parachute to open properly. We did a night drop on Salisbury Plain where we lost twenty men because the plane came in too low.

All this time I kept in touch with Janet by letter and, on occasional leave, I would go back to Cupar, a ten hour train journey, usually with standing room only. Fortunately, Peterborough was on the main Edinburgh-London line which made the journey reasonably straightforward. My relationship with Janet was now becoming serious. Marriage was on the horizon.

Our final practice drop was the whole Brigade together, 1,700 men, onto Salisbury Plain, in full battle gear, with heavy machine guns, mortars etc in a kit bag attached by a rope to our belts. When you jumped you had to push your weaponry out first and follow, dropping at a higher speed than before because of the weight, but the kit bag would hit the ground first and cushion your landing to some degree.

During the last drop I also had grenades strapped to my belt. As I was trying to clear a fence my foot got caught and I fell forward on the grenade, quite a painful experience. I had to be strapped up to go into battle with two broken ribs. On top of that, the drop was followed by a sixteen mile trek to Devizes in full battle gear. I was in agony but I couldn't let my boys down.

18
Katyn and Gibraltar

Two things weighed heavily in the minds and hearts of the Poles during this time: the discovery of Katyn and the mysterious death of General Sikorski.

The discovery of the mass graves at Katyn during the German withdrawal from the Soviet Union in the spring of 1943 was obviously a gift to Goebbels' propaganda machine. The Polish government-in-exile had been making continual representations to the Soviet government as to the whereabouts of the 17,000 Polish officers who had disappeared. Stalin said they were on the islands of Novaya Zemlya and various other remote places north of Vorkuta in the Arctic Circle, so it would be difficult to locate them and even more difficult to release them etc etc.

Then Katyn was discovered. Mass graves, corpses with bullets in the back of the head, the hallmark of the NKVD. The Germans immediately got an international commission to examine the graves. They confirmed the atrocity as Russian. Naturally, Goebbels made his point to the Poles and to the West. Just look at who your ally is! Look what he does to POWs!

Strictly speaking, the officers were not POWs. The USSR didn't declare war on Poland in 1939. It simply moved in under the ruse of 'protecting' the Ukrainian population in eastern Poland. In that sense Katyn was not a war crime but it was certainly a crime against humanity. At the end of the war the USSR was exempt from war crimes, a position re-affirmed with regard to the Russian Federation at the collapse of communism.

Naturally, Sikorski's government asked for an explanation. Stalin's reply was to point the finger at the Germans and break off diplomatic relations with the Polish government in London, saying the Polish army in the West was no longer regarded as an ally of the USSR. We knew then that Stalin would hold to the so-called Curzon line and that our return to eastern Poland would be virtually impossible.

Katyn was no surprise. We'd long suspected it. It underlined the duplicity of the Soviet government in everything it did; and the craven responses of both Churchill and Roosevelt reinforced a certain sense of betrayal. A deep bitterness about Katyn will remain with Poles for a long time to come. You have to remember that these officers were not regular soldiers, they were fighting the Germans in the west, but reservists. As such, they were mostly civilian professionals: lawyers, teachers, lecturers, doctors and so forth. In short, they were the intelligentsia of the country and that is why Stalin had them all murdered. This was, and remains, an unforgiveable crime.

The inquiry into the death of General Sikorski in a plane crash off Gibraltar in July 1943 declared it was an accident. Polish opinion varied, though. Trusting in our allies, the majority of the Poles in uniform probably accepted the official line. Today, in modern Poland very few people believe the accident theory.

Certain questions remain unanswered. What was the Russian spy Kim Philby doing in Gibraltar at the time? Why was the pilot Czech and not Polish? Why was he the only one wearing a Mae West (and, therefore, the only one to survive)?

The removal of Sikorski certainly suited Stalin. The status and influence of the Polish government decreased as a result. There was no other heavyweight to take his place.

I have an open mind about Sikorski's death, having yet to be convinced either way. At the time, however, it was another signal that a return to Poland was becoming virtually impossible.

In the spring of 1943 the Russian army had entered my home town, Luck. I knew that if I had stayed in the USSR and joined General Berling's army I'd have been home by now. But would I have survived the Eastern Front? And would I have survived in post-war Stalinist Poland? (Incidentally, some of my friends who were not deported to Siberia did survive, notably my first schooltime sweetheart Nina Slazkiewicz and my good friend Zbygniew Wincentak, who actually became an Air Force General and Deputy Chief of Air Staff in post-war Communist Poland).

19
Into Arnhem

Allied paratroopers being dropped over Arnhem. September 1944.

When the Normandy invasion began on June 6th 1944 we were put on immediate stand-by. Our British brothers in the 6th Airborne Division took part in the invasion and we knew our time was coming. We had been equipped with better arms: American automatic repeating rifles, Canadian Browning revolvers and the famous PIAT anti-tank rocket launchers. This certainly boosted our confidence.

The waiting game was hard and we tried to relieve the boredom by visiting the pubs and clubs in Peterborough, a somewhat more exciting place than Pittenweem or Cupar back in Fife.

In early August the Americans broke out of Normandy and were advancing rapidly through France. We were put on red alert. The hour was near. In early September the British reached the Albert

Canal in Belgium. We were briefed that an airborne assault was to take place involving the American 82nd and 101st Airborne Divisions and the British 1st Division with ourselves in support.

The 101st was to capture the river crossing at Graves then move on to Eindhoven. The 82nd would capture the bridge over the Waal at Nijmegen. The British 1st Division would be dropped on the northern side of the Rhine, the Polish on the southern side. The British were to capture the bridge and the Poles to cross it and advance over the German border. This would allow the British army to cross the three bridges, by-pass the Siegfried Line and advance into the Ruhr. The war would be over before Christmas.

The special objective for my company was to capture the military radio installation south of the Rhine and break the German line of communication with Berlin. At the briefing we were told that the allies had total command of the air and that the German army would be a rag-tag outfit made up of young men, the rawest of recruits, and older conscripts beyond the normal serving age. Resistance would, therefore, be minimal. The briefing was totally, absolutely and completely wrong.

The Dutch Resistance (as we later learned) had informed London that there were two SS Panzer Divisions in the area. Secondly, and unknown to the allies, Arnhem was now the HQ of Marshall Walter Model, a very experienced veteran of the Eastern Front and a Nazi fanatic, who had been sent there to bolster defences. Thirdly, there weren't enough aircraft to drop us all in one day, so the operation would be spread over three days, obviously leaving the lead troops in a very vulnerable situation.

The parachute regiments would go in on the first day. The landing brigade and heavy weapons (including the anti-tank artillery) would be dropped on the second day and the Poles would go in on the third. The armoured divisions would supposedly reach us within twenty-four hours, travelling sixty kilometres on the single road. The reality, of course, was different.

The Americans who were dropped at Graves had great difficulties getting through and were delayed for forty-eight hours. The British were still south of Eindhoven, sixty kilometres from their objective. The American 82nd Division captured the bridge at Nijmegen intact but the 1st Airborne Division were scattered and had several miles to march to reach their objective. They arrived and held the northern end of the road bridge but the 10th SS Panzer Division advanced to the southern end of the bridge while the 9th SS Panzer Division attacked the British from the north. The lightly armed parachute regiments had no chance against these heavily armed SS divisions.

To compound matters, in the heat of the battle a British officer who had landed by glider on the north side left his briefcase on the glider and it fell into enemy hands. It contained the Allied battle plans. So, unknown to us, the Germans knew the exact details of our assault plans. This was carelessness on the part of the officer concerned. He had to get out of the glider quickly in the middle of a battle, but to leave the plans was unforgiveable.

By the second day, the Germans were fully prepared. A lot of the gliders, now sitting ducks, were taken out before they could land. Of the twenty-four anti-tank guns being flown in on these gliders, intended for use by the Polish Brigade, two-thirds were either shot down or landed in such haste that they were rendered useless. They would have been invaluable against the Panzers.

Because of a lack of aircraft, the Polish field artillery was embarked on ships. We were effectively without most of the anti-tank guns and field artillery and had to rely almost totally on light weapons: three inch mortars and PIAT anti-tank launchers.

To make matters worse, the British commander of the 1st Airborne Division, General Urquhart, was separated from the Division, a situation exacerbated by radio sets which weren't working properly. He had no communication with his forces to begin with. It was only thanks to the Polish Division's signal unit who were with him and whose radios were working that he managed to establish communications with the Division.

Getting into Arnhem was also a long, drawn out process. The night before the drop we collected our ammunition and spent a sleepless night awaiting battle. Fears flooded in: fear your parachute might not open properly, fear you might let your colleagues down and, of course, fear of death itself. Some wrote last letters to their sweethearts, just in case. No one slept. Some prayed. Some were deep in thought. Mass was said. Guns were cleaned and oiled. Ammunition was checked. We were all carrying as much live ammunition as we could. I myself took three or four extra sten-gun magazines and lots of grenades.

In the morning we were taken to the airfields by lorry. Collecting and checking our parachutes, we hoped the WRAF girls had done a good job. We disembarked in the front of the aircraft, twenty people by each plane. As Deputy Platoon Commander I had to check the aircraft was properly equipped with weapons and that the static line for our parachutes was in working order. Then, just before we boarded the aircraft, the order came through that the operation had been delayed twenty-four hours because of fog.

Going back to the parachute stores and then to barracks we were full of mixed feelings. There was relief that we were certainly going to live for another twenty-four hours, but disappointment that we could not take part and support our colleagues in battle.

Then we received a briefing that, due to the presence of significant German forces in our dropping zone, we would now be dropped five kilometres to the west of the bridge, opposite where the British forces were establishing a defence perimeter. The idea was that we would capture the ferry at Driel and cross the river to support our British colleagues. Had we actually been dropped in the original zone we would have landed right on top of the 10th SS Panzer Division. That would have meant annihilation.

Next day we followed the same procedure with the same results: the operation was delayed because of fog over Holland. By this time the British paratroopers were in a desperate position totally surrounded by the Germans while the British ground attack had been halted at Nijmegen, the road having been cut in two places by German forces. Unknown to us at the time the whole plan was already in ruins and the operation was bound to fail.

On the third day, we managed to embark and take off. We sat on our allocated benches, ten soldiers on each side of the plane (a DC3 Dakota transport). There were 114 planes in all, carrying around 1700 soldiers. At last we were on our way to confront our enemy.

As I was going to jump last I was sitting nearest the pilot. On reaching cruising height I went up to the flight deck just to see what was what. We were above the clouds and being escorted by fighter planes, some of them Polish. In the cabin there was total silence.

After about half-an-hour we began a slow descent, presumably on reaching the Belgian coast. One of the crew emerged from the flight deck dressed in helmet and anti-bullet vest. He went to the door and opened it. His job was to stand there to help us step out in all our heavy battle gear.

The red light came on. We stood and hooked our parachute opening lines to a steel line running along the roof of the cabin. This was it. No more simulations. We were going to meet the aggressor – to take revenge for the rape and destruction of our country. We were going in.

The green light. One by one we stepped into the night air. As I was the last to jump I had to push hard, which I did with my head down so that I didn't even notice I was through the door until the cold air hit me. Three seconds. I heard the 'pop' of my parachute opening. Now the training took over. Thirty seconds to landing. I grabbed the front harness of the 'chute to control the swing, put my feet together with knees slightly bent. I could hear some shooting below. The welcoming party. We hadn't been expecting flowers.

The Germans knew what we wanted to do to them. Then, a heavy bump. I was on the ground. It took a few seconds to collect myself and assess the situation.

I could see members of my platoon running to the assembly area and could hear sporadic firing. Medics were running by with stretchers. They were going to have a busy night of it but I was too excited to have any fear. The adrenalin was pumping and I had to concentrate on the task in hand. But what was the task in hand?

Well, half-way over the Channel one of the American master navigators had decided to turn back because of the fog, so a third of the Division was returned to base although, as it turned out, there was no fog over Arnhem. Around one thousand of us were now landed and the Germans immediately threw five SS Battalions supported by tanks against us. So much for them being old men and little boys! The task in hand was to defend ourselves, to defend our position as best we could. The Germans were desperate to dislodge us because we now held the only escape route for the British 1st Airborne Division.

Unfortunately, the ferry we had hoped to commandeer had been sunk by the Germans and there were very few boats available when General Urquhart ordered the Polish Parachute Brigade to cross to the north side which the British were still holding. Some two hundred did cross under severe fire. The Germans were holding hilltop positions and bombarding us with heavy artillery and mortars alongside incessant machine-gun fire. A lot of the Poles were killed trying to cross in these boats. Our tragedy was we had nothing to answer their heavy weapons fire with, particularly their multi-barrel heavy mortars.

We had recognition squares with which to make markers for our transport planes to drop weapons, food and ammunition. Hearing a plane in the distance and assuming we had 'total command of the air' as our briefing stated, I ordered my men to make crosses on the ground with the squares. Then, to my horror, I saw a Messerschmitt 109 overhead. We lost two men before we could all scramble into a foxhole. Everything that could go wrong was going wrong.

The briefing had been wrong and planning was now being done on the hoof. We weren't out to capture anything now. We were simply dug in to defend ourselves; and from their vantage points the Germans could see our every move.

We occupied our defensive positions in Driel on the south bank of the Rhine opposite the defensive perimeter of our British brothers of the 1st Airborne Division. We managed to repulse the SS Division with support from the Guards' Division reconnaissance tanks which reached us before withdrawing south.

Our orders were to hold the south bank to allow the British to cross. On the night of 25th-26th September the orders came to cross the Rhine. Thanks to our holding our position around 2,000 British troops managed to cross the Rhine – though in all, some 10,000 would be captured and/or killed.

Of those who couldn't get across some did manage to evade capture and were hidden by local Dutch people so that during the next months a few hundred more would trickle over the river. A friend of mine, Stanley Kulik, hid out in a farmhouse for a month before crossing to safety with the help of the Dutch resistance.

A makeshift field hospital was also set up at Driel with help from the local people. I remember in particular the courageous efforts of Cora Baltussen who was to remain a loyal friend of the Polish paratroopers until her death in 2005. The brave Dutch people of Arnhem-Driel risked their lives tending to our wounded.

On the 26th we were ordered ten kilometres south to defend other bridges. A week or two later we were withdrawn by sea to England. Of the 1,500 Poles dropped one third were now dead, wounded or captured.

The devastating moment, psychologically, was the return, out of the heat of battle, to barracks to see the empty beds neatly made up waiting for those who would not return. What were we to do with their personal effects? We couldn't return them to Poland, to their families, things being what they were. Some were possessions that had travelled from Poland to Siberia and half-way round the world. Whatever happened to them, once we'd gathered them together, I never knew.

A friend of mine, Eddie Kostrz, had had a premonition. During the battle he said to me 'I have a feeling I won't be going back'. He gave me the address of his sweetheart in Cupar, Fife, and asked me to visit her and tell her to dispose of his personal effects as she saw fit.

'Come on, Eddie! It won't come to that! We'll get through!' I said.

But he was killed.

We got immediate leave after battle and I went to visit my Scottish sweetheart Janet who also lived in Cupar. I went to see Eddie's girl too. She had, it turned out, already received a letter from Eddie telling her he wouldn't be coming back. People do have such premonitions.

Another friend of mine, Kazik Kubrycht, came a cropper in an entirely different manner. When we had crossed the Channel and were flying over Belgium on the way into Arnhem the pilot told us to hook up in case of attack from Messerschmitts. As first in line to jump Kazik was standing by the open door when the plane hit a pocket of turbulence and he fell out. He broke his ankle on landing and was taken to a British field hospital before being sent back to England, having missed the whole shindig. To this day Kazik remains the only person I know to have fallen out of a plane!

For sixty years the role of the Polish Brigade at Arnhem has been badly under-estimated by historians and badly under-played by the British military establishment. In fact, General Browning, who was in overall command, blamed the Poles for his mistakes by saying we weren't battle ready. That added insult to injury. The truth is, if Montgomery and Browning had been German they would have been retired. If they had been Russian they would have been shot.

In 2006, the Dutch government awarded the Polish Brigade the highest military decoration and the British veterans of the 1st Airborne Division erected a monument to General Sosabowski, the officer commanding the Polish Brigade, at the town of Driel. This was worth more to us, as veterans, than even Queen Beatrix's award because it was done by the colleagues we had rescued.

Unlike Browning, Sosabowski had appraised the opposition correctly. He had fought the Germans during the invasion in 1939. He knew their mentality and he knew they'd defend their borders to the last man. At the senior briefing sessions he kept asking bluntly 'And what about the Germans?' He was told there would only be a token force. He didn't believe that.

I saw a lot of friends wounded at Arnhem and a few of them killed. In my mind I sometimes picture myself there as if I were another person, someone in a book I've read or a film I've seen. It's a psychological mechanism which helps you cope. Arnhem haunted me for years. It was impossible to suppress because people you met would keep asking you about it.

The memories were also mixed with disappointment, perhaps even depression, that we hadn't actually defeated the Germans there. That was bitter, knowing what Poland was still suffering.

The media treated us as heroes at the time but we didn't feel like heroes. The whole escapade was dressed up romantically. But the reality was that Arnhem was a major defeat and a propaganda coup for Hitler.

A lot of people talk about going through the 'pain barrier' in extreme circumstances like those of a battlefield. For me it was more about the 'fear barrier', which I went through three times: during NKVD interrogation, during the Gulag and during the Battle of Arnhem. In each case I was scared out of my wits, scared of being hurt, scared of dying and scared of fear itself. But there comes a point of acceptance. During interrogation you accept that what they're doing to you can't really get any worse and you see death as a release. Interrogation prepared me for the Gulag and both these experiences prepared me for Arnhem. Seeing people die in the Gulag and next to you in battle you face the same fear again. You say 'there's my colleague lying dead, it's actually happening!' Then you get angry and want to kill those who are killing your friends. That's when you get 'heroic'. Even the wounded will fight like hell because they are so angry. And this is the danger mark, because now you can become barbaric. You can lose control completely. You have to understand these emotions before you make armchair judgements of what people do in a war situation.

When my friend Eddie Kostrz was killed beside me at Arnhem I was flooded with anger. I didn't care about the bullets flying around me. I just wanted revenge.

Some people are suspicious of military tribunals examining alleged 'atrocities'. I am not. A tribunal looks at the 'atrocity' from a military viewpoint, not a societal or sociological one, and that to me is correct. You have to know it, to have been through it, to judge. That doesn't mean there are not sadistic people; but there again experienced army people are in a better position to judge, to distinguish between anger and sadism. War *is* cruel. That's not a cliché. And soldiers regret even justifiable actions.

Polish Paratroopers on the south side of the Rhine near Arnhem.

20
The Warsaw Rising

The official motto of the Polish Parachute Brigade was 'The Shortest Way' - meaning the shortest way back to Poland. This certainly implied that when the Soviet Army invaded Poland the Brigade would be dropped in to support the *Armia Krajowa,* the Home Army, which we always knew would rise. This is what we were fighting for and what we wanted to do. It had always been part of General Sikorski's plan. But Churchill, Montgomery and others thwarted it because they wanted us on the western front.

Polish Special Operations Executive (SOE) officers were dropped into Poland throughout the war to help organise the Home Army. These officers were trained alongside us at Upper Largo and Ringway and we knew some of them went in and out of Poland on Lysander aircraft. Indeed, in one of those strange co-incidences, I met up with the Lysander pilot, Group Captain Hockey, after the war. He was my boss at S. Smith & Sons in Cheltenham and he confirmed these flights took place with some regularity.

When the Soviets advanced into Poland in the summer of 1944 they captured Lublin and set up a communist puppet government, the so-called 'Lublin Government'. On 1st August, with Berling's troops already in Praga on the outskirts of Warsaw proper, the Warsaw Rising began, encouraged by Soviet radio broadcasts. The Home Army had no choice but to rise. Stalin had them in a lose-lose situation. If they didn't rise, he could accuse them of cowardice, of being German sympathisers, *bourgeois* reactionaries and so on. They fully expected Soviet help but Stalin had no intention of coming to their aid. He wanted them slaughtered. If the Germans could do it for him so much the better. It would save him having to organise another Katyn. They rose and he left them to their fate.

It was almost as if the Nazi-Soviet Pact was still in effect. Between them, Germany and Russia were still intent on the destruction of Poland. The Germans withdrew initially, expecting Soviet troops to come flooding over the Vistula. When they didn't, and Hitler realised what Stalin was up to, the Germans regrouped. Members of the Home Army, half a million strong, were hiding out in the dense forests surrounding Warsaw waiting for the moment when the Soviet Army would move into the city. But that never happened until it was too late for them to be of any help.

By this time the Germans must have known they were beating a long retreat to ultimate defeat, but still they set out not only to defeat the Rising but also to raze the grand old city, the capital of Poland, to the ground. It was revenge, pure and simple. They knew they were going to be destroyed, so they decided to take Warsaw with them.

Part of the German army in Warsaw was made up of Ukrainian units under the command of General Vlasov and they were particularly cruel, not to say barbaric, in seeking what they saw as revenge for the traditional Polish domination of western Ukraine. There were, as the eminent historian Norman Davies has pointed out, many more Ukrainians fighting against the *Wehrmacht* than for it; and given the terrible atrocities eastern Ukraine suffered under prewar Soviet rule it is perhaps surprising how few Ukrainians actually joined the Nazis. But here, I believe, Hitler missed a trick. Blinded by his own propaganda that all Slavs were slaves, he never took too kindly to our neighbours. Had he been more civilised in his treatment of the Ukrainians many more might well have joined Vlasov's ranks; many more might have helped him fight the Soviets from the start. That said, a Home Army soldier or Warsaw civilian had a much better chance of retaining life and limb in 1944 if captured by a German rather than by a Ukrainian soldier.

During the time of the six week long Rising we had liaison officers from the British 1st Airborne Division stationed with us and I remember discussing the plight of Warsaw with one of them.

'I know you what you're feeling, Joe, what you're all feeling' he said. 'I know you want to fight in your own country and we'd go with you if the orders came. But in the present circumstances it's just not viable. Stalin won't allow our planes to land or re-fuel on Soviet controlled territory and, anyway, the losses would be enormous.'

Some of my Polish colleagues argued that we should still go come what may, but I think it was the right decision. We would have needed two hundred planes for an initial airlift and these planes were desperately needed on the Western Front. To drop arms, ammunition and food in would have meant two or three such flights. The Americans would just not risk so many planes and so many pilots.

Poland was fast becoming a dream. Our route to Poland was being blocked at every turn. Since the vast majority of our Brigade had come through the Gulag few of us would ever consider returning to a Soviet-controlled Poland. We remained absolutely loyal to our allies and to the Polish Government in London. We still didn't know that Poland had been sacrificed to Stalin at Tehran in 1943, a decision ratified finally at Yalta in 1945 when it became public. So, throughout the Battle of Arnhem, Warsaw was very much on our minds and there was, as there had been since 1939, a degree of hatred in our fighting souls. The Germans knew that with us it would always be a different fight.

At the end of the Rising Hitler ordered the complete destruction of Warsaw. The entire city was blown to smithereens. The population was killed, dispersed or sent into slave labour or the concentration camps. It took years and years of patient, and often voluntary, effort to rebuild the city. That Warsaw is today once again a dynamic capital is proof that you cannot defeat the human spirit.

Warsaw left its mark on all of us and when I began revisiting the city in the 1990s people still said 'Why didn't you come? We were waiting for you'. There is still a bitterness about this in the city, a bitterness which is only beginning to recede with the third post-war generation. Only the grandchildren can begin to grasp the historical perspective.

The Rising was portrayed by many in the West at the time and in the aftermath of the War as romantic folly. It wasn't. Warsaw did not bring destruction on itself. *Generalissimo* Stalin did that with the help of Corporal Hitler. And when I hear, as I have heard so often in recent years, that Britain should apologise for the bombing of Dresden, it seems incomprehensible to me. In the order of magnitude what happened to Dresden is simply not comparable to the barbarism that wiped Warsaw from the face of the earth.

Warsaw, January 1945

21
From Arnhem to Germany

Post-Arnhem our Brigade had to be re-formed because of the huge losses. But where were we to get new Polish recruits? Every fit and able Pole was already in uniform. Some of them, though, had been in *Wehrmacht* uniforms and were now POWs.

These were conscripts, for the most part from Pomerania, Silesia and Poznan, areas of western Poland which had been incorporated into the Reich. Some Poles in these areas had been granted, or had applied for, *Volkdeutsch* status. As German citizens they were thus liable to conscription, especially in the later stages of the War because of the heavy *Wehrmacht* losses. As POWs they were now being given a choice: join the Polish forces or be sent to a POW camp.

Most of these men were not willing German soldiers, as Churchill had recognised from the outset, but there was still resentment among the free Poles over their recruitment as only weeks previously they had been wearing German uniforms.

Most of us were uneasy about this and I was in a dilemma. My battalion had only remnants left and I was given a choice: remain in Peterborough and join the 1st Battalion or return to Upper Largo and train up some of these ex-*Wehrmacht* soldiers. I felt I couldn't honestly put my heart into the latter and so was given command of a heavy machine-gun platoon. I was sent for three months training to Netheravon in Wiltshire where I was one of very few 'Johnny Foreigners'. This helped my English greatly; though, as I was befriended by an Irishman called Docherty, I almost developed a Polish-Ulster accent. I was also beginning to understand British people better.

We were training on Salisbury Plain with live ammunition and watching the 6th Airborne Division's preparations for their drop into the Rhineland in March 1945. This time they would be dropped a few kilometres ahead of the land forces and not sixty as at Arnhem. I recall seeing a tragic accident when one of their men got his parachute caught on the rear wheel of a plane. As the plane circled he fought unsuccessfully to get it free. Eventually, the plane swooped as low as it could over a reservoir and he released himself but did not survive impact with the water.

In April I was recalled to take charge of the heavy machine-gun platoon as a Second Lieutenant. My new commanding officer, Lieutenant Szaflarski, was something of a character. Tall, dark-haired and with a booming voice, he'd been captured in 1939 but had escaped from the POW camp in Germany and made his way to Britain via France, Spain and Portugal. He was a terrific bridge player and after the war represented Britain in international tournaments playing alongside people like Omar Sharif.

Besides improving my bridge of an evening I had to train my platoon in their new heavy weaponry and get them battle-ready. Then, in May 1945, we were landed by sea at Ostend and dispatched to eliminate a pocket of German resistance around Dunkirk. We were eager and ready for them. But as soon as we arrived the Germans surrendered. It may be that they knew they were fighting a losing battle but I think they surrendered because they knew we were Poles and that we would make it one hell of a fight if they decided to engage with us.

After this we were sent on to West Germany, first to Emmerich near the Dutch border and then to Osnabrück to take up our occupational duties. It was strange to be on German soil. I was glad the war was over and very glad Germany was defeated; but I felt a degree of sorrow for the German population. You could see that people were confused and didn't really know where to go. The streets were full of refugees carrying everything they had in meagre bundles, looking for relatives, looking for a home, disorientated, lost. It was not dissimilar to Poland in 1939. A lot were very hungry and grateful even for a piece of chocolate.

Before we had left for Germany I kept asking myself what our attitude towards the German people would be, and what their attitude would be to us. We were both apprehensive and curious as to how the 'superior' German people would react to taking orders from us 'sub-human' Slavic people.

Be in no doubt, we hated the Germans for starting the war and for their utmost cruelty to our people throughout the it. We knew that the vast majority of Germans supported the Nazis and their policy of Polish conquest and subjugation and we knew the townspeople and villagers we had to police were very apprehensive of us. But they needn't have worried. In contrast to what they had done in Poland we were correct in our treatment of them. Perhaps they were surprised that 'inferior' Poles could behave towards them in a civilised way.

As we moved east towards Osnabrück and saw the devastation caused by the allied bombing our feelings of hate were replaced by pity and compassion. After a few months one could even detect occasional friendships between young Polish soldiers and young German girls – but never between young Polish soldiers and German men.

The streets were also teeming with slave labourers now free but far from home: French, Poles, Yugoslavians, Romanians, Greeks, Lithuanians and Latvians all trying to find their place in Europe and to re-establish their lives from virtually nothing.

So, from a fighting unit, we became 'policemen', making certain that the DP (Displaced Person) Camps were being properly run and vetted. We had to make sure that former Polish prisoners did not get out and take revenge. I don't think this was ever really going to happen but we had to be careful. We had to feed, clothe and care for them. There were masses of these camps all around Osnabrück and a lot of our soldiers found new Polish girlfriends.

We also had to smash up illegal stills for the making of schnapps in both camps and villages and to confiscate black market tobacco and so forth. As a highly trained soldier I hated these duties. But we had to maintain law and order and act almost as a social service.

There was a 'no fraternisation' policy between the allies and the civilian German population at this period for both medical and political reasons: fear of the spread of VD among the forces and, with anti-Nazi hunts still going on, the fear of fraternization with the enemy.

The war with Japan was still going on and there were plenty of rumours that we'd be sent to the Far East. As Poles we had no personal quarrel with Japan and no great enthusiasm for such a campaign, though no doubt we would have done as ordered. But because of the Atom Bomb and the Japanese surrender it never happened.

This compelled us to think about what we were going to do now. I knew my career in the army was as good as over. The idea of attacking the Soviets in Poland was unrealistic, though some Poles still entertained the idea. The USA and Britain just wouldn't wear it. Britain was war-weary and had no stomach to pursue Polish rights or hopes. Besides, it was totally bankrupt. Tehran, Yalta and Potsdam had decided our fate. Washington and Moscow had divided Europe and Britain had little or no leverage.

In fact, in economic terms, Britain and Poland, the two oldest Allies, were the two long-term losers from the war. While American money was used to boost the post-war West German and Japanese economies (the two 'losers' of the war). Britain remained, relatively speaking, in the doldrums, no longer a world power; and Stalin's gluttony swallowed Poland whole.

As Poles, we had three options: wait to be shipped to Poland; return to the UK; or go to whatever country would have us. The first option was a non-starter. I was aware that most of my colleagues from Eastern Poland weren't going back. Like them, I knew the communists too well. I wasn't for going back to a Soviet occupied Poland and, having by now a Scottish sweetheart and plans for marriage, I certainly wasn't going to expose my future wife to any unnecessary dangers.

Some of my colleagues did go back to Poland with their Scottish wives. In most cases, it ended in tears. On top of all the repression, the living conditions were desperate. Maybe a few hundred wives went but practically all returned and a great many had terrible difficulties getting their husbands back out. Whether all of these husbands did make it back I don't know.

What connections I now had were British, or Scottish, and the idea of emigrating to somewhere like Canada or Argentina had no appeal. At Battalion HQ I'd seen a notice saying that the UK Government was offering a grant to Polish soldiers to study at a university or polytechnic. Although my ambition at school had been to become a lawyer, I decided I wanted to do engineering because it was practical and post-war Britain would need engineers. Besides, engineering would not demand as sophisticated a knowledge of English as a humanities subject would.

My last weeks in Germany were a curious mixture of joy and sadness. Joy because the war was over and we had seen millions liberated from the German work camps and the extermination camps, people who could at least attempt to resume a normal life now their slavery was ended; sadness because Potsdam confirmed Yalta and we knew we wouldn't be going home. Stalin gave assurances that Poland would be 'free'. He even offered a deputy premiership to the Polish Prime Minister in exile, Mikolajczyk. In Germany we knew this 'democratic' set-up was exactly that, a set-up.

Then Stalin invited the Polish Underground Government in Warsaw to talks in Moscow. Seventeen of them went including the chairman, my old divisional commander in Iraq, General Okulicki, who had been dropped into Poland to replace General Bor Komorowski as commander in chief of the AK after the Warsaw Rising. These seventeen men were arrested in Moscow, put on show trial, and sentenced to prison where Okulicki was to die. Prime Minister Mikolajczyk did go to Poland but the CIA got him out before he was arrested.

This was the last act of the rape of our Poland. We felt totally alone and utterly betrayed: betrayed to the extent that we were not asked to take part in the 1945 Victory Parade in London – the final insult – because Stalin objected. We were Britain's first ally. Britain had gone to war so as not to appease one dictator and ended up by appeasing another. Maybe Churchill tried harder than his American colleagues, but Roosevelt would do nothing that would anger Stalin whom he seemed to like and trust.

The Soviet contribution to the war effort was very high – that could never be gainsaid – but the price the Allies had to pay for it was Poland. We had sacrificed all for freedom and now we were shunned. It was a Greek Tragedy on a macro-scale involving an entire nation.

22
Poles Apart in Portsmouth

In early September 1945 I was given leave of absence. I said goodbye to my friends in the Brigade. It was a sad moment. These were friends with whom I had shared so much throughout the long years of the war. But a new life beckoned.

I surrendered my personal weapons and ammunition to the Quartermaster, surrendered my platoon to my successor, reported to the Battalion Commander to thank him for his support and said a final farewell to my closest friends before embarking on the train to Ostend for the boat to England.

I was full of apprehension about how I'd cope in 'civvy street' and at a British college. I felt my English was good enough to cope with higher education and I knew that Janet would give me all the support she could. Nevertheless, although the war years had made me grow up quickly and take decisions, and responsibility, for myself, I was missing my parents badly and thinking about what advice they might give me.

When I arrived in London I reported to the Polish Government Department of Education who tested my knowledge of English and my general level of education to see if I was university material. There were two possibilities, Huddersfield or Portsmouth Technical College. I don't know that I'd ever heard of Huddersfield but I certainly knew of Portsmouth because of the naval base. So I reported there and was tested again by a Professor Benny, a Cambridge man, who said he'd be delighted to have me start that month.

Josef in Portsmouth, 1946

I received a grant of £30 per month, a pitiful amount by today's standards, but it was sufficient then and I could even save a bit. I found digs with a Mr and Mrs Fudge whose only son, a submariner, had been killed in action in the

Mediterranean. We got on extremely well. They treated me like another son and we kept in touch for a long time afterwards.

They took in two other lodgers, two WRENs around my age who both had boyfriends. I socialised with them as a group, going dancing and visiting towns along the south coast, and this did a lot to integrate me. I got to know English culture as I had Scottish.

Portsmouth Technical College, 1946

In post-war Britain I was confronted by a different culture and faced with the inevitable dilemma every forced migrant, every displaced person, has to deal with: should I hold on to my birth culture or adopt the host culture?

In one sense I had no choice. While knowing I would always be Polish to a greater or lesser degree I had to take conscious and positive steps to embrace Britain. There were only two or three Poles at the college and I was cut off from Polish organisations which were centred in London. Although my grant was sufficient I was hardly able to swan up to London every weekend. Nor did I want to.

I have to be honest and say that I was disappointed by what I was reading in the Polish expatriate press at the time and I didn't agree with the attitude of the London Polish scene which to my mind was factionalised and backward looking. In a real sense they were living in the past and not accepting change. It looked like they were creating a Polish ghetto centred on Ealing – a Polish mutual admiration society.

Practical engineering. Josef on far right.

I blame the Polish government-in-exile for neglecting the welfare of the ex-combatants. More could have been done to

August 1947.
Marriage. Kirkcaldy, Fife

insist that all Polish soldiers learned English and more time and resources should have been given to this.

The Polish government-in-exile encouraged the idea that we could still drive the Soviets out of Poland. This was wrong. It was naïve and totally out of touch with reality. It was this attitude that was responsible for many very able Poles working in menial jobs while supposedly waiting to return to Poland instead of forging ahead with a new life in Britain.

This attitude was coupled with resentment from British people who couldn't understand why we were not welcome back in the country we'd fought so hard for. Stalin was demanding our return, of course, but that's a different thing from a welcome!

If British citizens were naïve or misguided, one man who knew the score was the Labour Foreign Secretary, Ernest Bevin. He must have had access to intelligence reports from Stalinist Poland and must have known what was happening to the ex-combatants who had returned. Yet in Parliament he was encouraging us to go back.

Of the 2,000 members of the 1st Polish Parachute Brigade, about half went back. Most of these returnees were from western Poland, people who had experienced Nazi but not communist oppression. The vast majority of the eastern Poles, who knew first-hand what the communists were capable of, were not prepared to risk it.

Of those who did go back, practically every one of them was either arrested and deported to the Siberian Gulags or refused decent employment because they were deemed 'untrustworthy': they were full of 'Western ideas' and regarded as potential spies.

Take the case of Professor Podoski whose father, also a professor, had been Dean of Warsaw Polytechnic. The communist authorities lured him back by offering him the same position of Dean. Poland was short of intelligentsia, due to their mass murder by the Germans, and by the Soviets at Katyn.

For a few years Professor Podoski served as Dean before being arrested as a 'Western spy' with other former officers from the Free Polish Forces. He got eight years hard labour in the notorious Wronki Labour Camp, but was released after four or five. And this is what they did to someone they *wanted* back.

When he was pardoned, Professor Podoski bravely insisted it was not enough. Eventually, he did get total rehabilitation. His case, though, illustrates the Stalinist strategy: use them, then jail them. That was typical of the fate of many who returned.

During the four years I spent in Portsmouth I was not exposed to much anti-Polish sentiment – many other Poles got it a lot worse. This may have been partly because there were relatively few Poles in Portsmouth but I believe it was mainly because I had good English (something I had worked hard at).

Nothing annoys indigenous people more than hearing foreigners speaking their language badly or with strong accents. I think the indigenous feel threatened in some way by this. I also think a lot of racism today still stems from this kind of feeling.

A lot of my Brigade colleagues remained in Germany until 1947 when the Polish army was demobbed. By this time I had a two year advantage on them in terms of English language and knowledge of post-war Britain. Most were sent to de-mob camps, the Parachute Brigade to one in Inverary in Argyll. A lot of them emigrated: to Canada, Argentina, Australia and New Zealand. The Antipodes were popular because in North Africa the Poles had fought as part of the Australian Division and the Australian Government had facilitated Allied Polish Soviet relations after Moscow broke off relations with the Polish Government in London.

September 1957.
Durban Road, south east London
Jose's first day at school.

Sadly, then, the great Brigade dispersed around the world. I lost touch with close colleagues. Now I had to get on with a new life in Britain.

In 1947 I married Janet at the Catholic Church in Kirkcaldy. She was only nominally Protestant but still Kirk-oriented and would never have considered becoming a Catholic. I respected her for that. I was what you might call a doubting Catholic after my war and Gulag experiences – how could God allow such things? I stopped going to church around this time, but when our only daughter, Josephine (Jose), was born in Cheltenham in 1952 we had her christened. We left it to her to be confirmed which she was an adult.

I did witness a degree of anti-Catholic sentiment in Janet's family, though it was never directed at me. Her father and two of her brothers were in the Masons which, in Scottish terms, is often regarded as only one step away from the Orange Lodge. Scottish rural communities and mining towns like those in Fife were often staunchly Protestant in those days. But I as only aware of real animosity once, at an Old Firm game on a visit to one of Janet's brother's, Jimmy, in Glasgow in the mid-1950s. Jimmy was a Rangers man, naturally. We were late for the game and had gone into the 'wrong' end. A Rangers player was tripped. Jimmy shouted for a penalty and there was a bit of a hullabaloo around us. So we made a sharp exit to the other end of the ground. We actually ran! It seems funny in retrospect, but it was quite serious at the time.

After we were married, Janet joined me in Portsmouth and we stayed with the Fudges for a couple of weeks while looking for a flat for ourselves and Janice. We were to spend forty years together, mostly in England where Janet never lost her Fife accent. If anything, she became more Scottish the longer we were in England!

As jobs were not plentiful in Scotland we were not to return there until I retired in 1983, though we visited her parents and family once or twice a year, usually for summer holidays and sometimes for Christmas and New Year.

Janet held great New Year parties in England and she was a wonderful cook, introducing me to lamb which has never figured largely in Polish cuisine (partly, I think, because of associations with the Lamb of God).

In 1948 I received my naturalisation papers. I was now a British subject. Up until that point I'd had only an Alien Resident Permit (I had no passport whatsoever) and had to report to the police station every month to have it stamped by a plain clothes officer. Because of my NKVD experience I had no love of police or police procedures. I was always on edge during these visits and I bitterly resented them. On receiving my naturalisation papers, I took the Alien Permit and tore it to shreds.

23
A Displaced Person

I still often think of my old pal Freddie Sodowski, who kept popping up in my life. Saying a little about Freddie and our friendship might help illustrate what it is to be a 'displaced person'.

Freddie was a classmate at the *gymnasium* and, though not close pals, we shared a passion for sport. When I was put into the holding cell at St Brigitta's after I'd been sentenced, he was already there. Recognising me, he made room beside him so we could talk. Like all the Poles there he was in for AK activities.

We were put on the same transport and saw each other all the way to the final railhead in Siberia. There we parted into different slave camps.

It was not until over a year later when I was in Pahlevi in Persia (Iran) that I bumped into Freddie again. There, we were mates again for a few days before being dispatched to different units in Iraq. Once more Freddie disappeared from my life. Then, on boarding our French transport ship, *Isle de France*, I met Freddie again. He'd volunteered for the Polish Airforce.

After our round-the-world trip, when we disembarked at Gourock we were split up once more. It was not until after the war that we met again through a mutual friend. Freddie had been a fighter pilot and was now living in Sheffield. Over the years we visited and talked together and often played golf either in Sheffield or in my adopted home town of St Andrews.

Until his death in 2001 Freddie was the only person who provided continuity from my young life in Luck, through the Siberian Gulag to Iran/Iraq to post-war Britain. He played a great role in underlining my Polishness. I always felt where my roots were when I met with him. I felt I was meeting a brother.

We were very much 'displaced persons', for forty odd years cut off from our homeland, from our roots, by the Cold War. Keeping in touch with a childhood friend like this helped me maintain a unity in life, a through progression, a spiritual or psychological integrity. It helped keep me *me*.

24
A Travelling Man

After college I had a great deal of difficulty finding a job, partly because of prejudice, but also because there were now five million people discharged from the army looking for work. I remember I even wrote to ICI for a job as a simple process worker in spite of my degree. The reply said 'You are a highly qualified engineer and should be applying for a better job'.

My wife, being a canny Fifer, said I should write directly to George Isaacs, the Minister of Labour in the Attlee government.

'Aye', I thought 'That'll be right'. But *she* did. I don't know what she put in the letter – the whole idea was just laughable to me – but two weeks later there was a knock at the door and there stood a bowler-hatted gent with a briefcase.

'I'm looking for a Mr Josef Tarnowski' he said.

' I'm Mr Tarnowski', I replied.

'Well, sir, I have an order here from the Minister of Labour to find you a job today.'

And he did. That afternoon I had a job in the design office of the Decorative Paper Company in Gosport, Portsmouth. My industrial career had begun.

DPC was part of Brighter Homes which supplied wallpaper, paint and equipment to the domestic and industrial markets. My job was to design block prints, stencils and rollers, to semi-automate the process for 'hand'-painted items which the public paid more for. It was interesting electro-mechanical work but it was a narrow field. I wanted a bigger challenge with more electronics experience.

Applying more widely, I landed a job at S. Smith & Sons Aircraft Instruments Division in Cheltenham and became part of the team designing the first fully electronic automatic pilots; in short, the first auto-pilots for aeroplanes. It was a challenging job. Then I was suddenly shifted to the production control office which I accepted because it widened my experience, letting me see how designs were actually produced on the shop floor.

We were staying in a rented flat near Cheltenham Ladies College, a very 'select' area full of retired army types 'taking the waters' (which were actually pretty foul to drink). Then Janet fell pregnant. Children weren't allowed in our rented accommodation and we couldn't afford to buy, so we applied for and got a company house on a nearby estate. With thousands of others I cycled to work every day in our modern, single-storey factory.

I was enjoying the work but was soon to learn that my progress in the company would be blocked by H.M. Government because of my Polish background.

The company was producing mechanisms for rockets under an MOD contract. The Cold War was in its first big freeze and because my parents were still living in Poland it was 'thought' I might be got at by the Soviets. I was a little surprised at the government's attitude but realised that this was probably why I'd been moved from design to production. It was time to look for another job.

I applied to the BBC for a job as an engineer. This time the Cold War factor came into play in a different way. They had no engineering jobs but because I spoke Russian they offered me work with the World Service. I'd decided to keep well out of politics at the end of the war and I'd met too many boring British 'spies' who'd worked at Cheltenham GCHQ in my time in that Spa town, so I turned the offer down.

I also had an interview with Hoover in Glasgow but, as they too were working on MOD contracts and as it would only be a matter of time before I was blocked, I turned them down too. At an interview with English Electric in Luton I was actually told by my interviewer: 'I really shouldn't be telling you this but, although I'd like to employ you, you will be knocked back by the MOD'.

So although I was a British subject I was still being impeded by my Polish roots. In this instance, the impediment was political – the Cold War. On two later occasions it would be racial.

In 1970, for example, apropos of nothing, my boss suddenly apologised for not appointing me chief engineer. He never said why. I hadn't applied. It was just one of these appointments that were 'made'. I think there had been a feeling that British employees would have complained at a Pole being promoted. Later still, in 1982-3 one of my bosses actually said to me out of the blue 'If you were British, Joe, you'd be a divisional manager at least'.

I decided, then, to look for work in companies with no MOD connections. In 1954 I landed a job with Standard Telephone and Cable, part of the multi-national ITT, one of the biggest companies in the world. We moved to London and I joined the Transmissions Division at North Woolwich as a design engineer on landline telephone systems. The building we worked in was a pre-war factory with an entrance not dissimilar to a public toilet's. It was also next to the River Thames which was filthy and stank all summer. There was no air conditioning and you couldn't open the windows for fear of that deadly stench.

The work was boring but the people were friendly and three colleagues who were to become good friends, Charlie Abrahams, Ernie Miller and Stan Plater, introduced me to how STC worked. It was very conservative. Everyone addressed everyone else as 'Mr' and there was a comically rigid hierarchy. You might begin in an open plan kind of office with just a desk among others, then you'd move up to an office which had a doorway but no door, then to an office with a door, then to one with a door and a carpet and so on. That was just the way it was.

I was working on a telephone system to put twelve speech channels through one copper wire. It was cutting-edge technology, though it may seem laughable now. When I left ITT in 1983 we were transmitting 10,000 speech channels through one optical fibre no thicker than a hair.

Things began to change rapidly around 1960 when the boss of ITT, Harold S Geneen, decided to build a new, modern factory in Basildon. I was transferred around this time to microwave systems which were seen as the future for the company, especially with the rapid development of television. This was a significant move for me. I was now a senior designer and world travel beckoned.

My first assignment was to tour our research laboratories in Europe, in Paris, Zurich and Brussels, all major ITT centres, to see how they might help develop the new microwave system which, among other things, would make telephones accessible to ordinary people. It was fantastic work designing transmitters and antennae and involving civil, electronic and mechanical engineering skills all in the one job.

Under Geneen the ethos and culture of STC, the subsidiary company, began to change, becoming more like those of its American parent company. The move to a bright, new modern factory effected a move to meritocracy. The old boy network no longer counted, which was great news for me. You were now given opportunities irrespective of your nationality or background. I was to work with giants in the industry like Alec Reeves, inventor of the pulse code modulation (PCM) system, the basis of digital transmission, and Professor Charlie Kao, inventor of optical transmissions for telephone systems, who went on to become Principal of the University of Hong Kong. As an engineer I couldn't have wished for a better company to work for.

That first European tour involved travel by train, plane and helicopter. It began a period of frequent travel in Europe as the research and development of the microwave transmission system was so enormous and complex. Microwave was the child of the Radar system which had given the UK the lead in the field at the time.

I was now to become a regular visitor to West Germany, to the town of Pforzheim near Stuttgart.

On my first visit I was both apprehensive and curious. My new German colleagues would know from my name and my Polish-accented German what my roots were and, given my age, that I must have fought in the war. But they chose to ignore this. To all intents and purposes I would be treated as an Englishman. It saved them a lot of embarrassment. They didn't need to apologise for any atrocities committed in Poland and, in fact, went out of their way to make me welcome.

When I got off the train in Pforzheim that first time I took a taxi to my hotel. The first place I remember passing was a warehouse which was distributing gas cylinders. It sent a shudder through me. Was Zyklon B (the poison gas used in Auschwitz and other death camps) distributed from here?

I was greeted by our local manager, Dr Laaff, who explained that the town, which was in the middle of the Black Forest, had traditionally produced gold jewellery.

'Yet we were bombed by a hundred bombers during the war. Why?' he asked, seemingly bemused.

What he forgot to say – and what he must have known being an engineer himself – was that Pforzheim was also a telex manufacturing centre producing the Ultra coding system for the German army. The Ultra is perhaps better known as the famous Enigma machine used by the *Wehrmacht* for sending coded messages.

For me this was a curious coincidence. Not just because I was Polish and it was Polish engineers who actually decoded the machine in Warsaw as early as 1926 before offering it to French intelligence (who weren't interested) and then to Britain who eventually deciphered it in full at Bletchley Park. Now, I knew Bletchley Park because it became an STC facility after the war and I used to visit it for company meetings, conferences and so on. I used to smile ironically as I crossed the threshold. To Dr Laaff I just grinned broadly. (Much later I was to meet Irene Thornton, the translator of Professor Garlinski's book on Enigma, in her role as the original secretary of the Scottish Polish Society. Her Polish husband had been dropped into Poland to help organise the Home Army. The Soviets shot him).

I was given a car with a driver and at weekends I could tour the Black Forest area, being put up in the best of hotels. I was certainly being well-treated but I couldn't erase thoughts about the war and what the German occupation had done to Poland. There was even a little village near Pforzheim called Birkenau which made me think constantly of that other Birkenau.

There were reminders all the time, including a lot of bombed out buildings and German war-disabled. But American dollars were also pouring in and there were brand new factories going up all over the place. The German economy was rebuilding at a much faster pace than the British. On every visit there was more and more evidence of affluence compared to the UK. There was also more dynamism: more fresh ideas in organised production while, back in Britain, operation systems were often stuck in war-time mode and the operators fixed in a pre-war mentality. I was really learning about electrical engineering now and I was also visiting Belgium a lot. Here the economy was expanding almost as fast as in West Germany.

I was also learning about capitalism, a curious beast, indeed. An old-fashioned Marxist (and I'm not really sure there is any other kind) might argue that capitalism inevitably leads to war in its drive for materials and markets. Obviously this happens, or has happened, but to set it down as some kind of 'law of history' is nonsense. Powerful companies can be directed by, or work in cahoots with, governments for political ends. You have only to think of the role the East India Company played in the expansion of the British Empire or, indeed, the role of my company ITT in undermining Salvador Allende's socialist government in Chile in the service of the dollar. Equally, multinational companies can control or corrupt governments as has happened in many third world countries.

But the only war that business is really interested in is economic. And here I use the word 'war' as a metaphor. The drive for profit, as part of a larger and more complex matrix, can lead to conflict: to literal war. But conversely, even in a war-torn world, companies will seek to go about their business peacefully in the drive for profit.

I can think of no better illustration of this than ITT itself. During World War II such was the size and power of the company that European executives regularly met and reported in neutral Switzerland. These executives came from Britain and Germany, as well as from Belgium, Sweden, the USA and neutral Spain. Capitalism has no borders and it doesn't necessarily recognise military fronts.

I am of a liberal cast of mind and I hold no brief for unfettered capitalism, for laissez-faire economics or old-fashioned Thatcherism. A balance must always be struck between profits and people and, in terms of health, wealth, safety and freedom, people must always come first.

The distribution of wealth (it is patently not a *re*-distribution as Marx would have it), in this sphere, is the role of government (as well as of private individuals). But you cannot distribute what has not been created. Wealth creation is the role of free enterprise; an enterprise whose freedom should never be outwith civilised limits. Profit will always be sought for its own sake but a balance of the surplus must serve society as a whole. Finding that balance is, of course, no easy matter.

Capitalism was certainly turning my life around in strange ways. A Pole recognised by his German colleagues only as an 'Englishman' was now asked to become a German representative. This happened during a busy period in Pforzheim when Dr Laaff couldn't spare one of his own engineers to go to a meeting of international colleagues. So I was asked to represent Germany! In a way it was quite satisfying, given that less than twenty years previously these same people would have regarded me as an *untermensch*, a 'sub-human'.

Talk of the war remained taboo. As a matter of principle I never broached the subject and neither did they. In any case, we were all developing friendships across Europe, from Sweden and Norway to Spain and Italy. We were international teams and friendships overruled national boundaries and prejudices. That was certainly a very positive thing and one of the most enjoyable aspects of working for ITT. In some cases personal friendships became family ones. My Belgian colleague Nick Artz had a daughter the same age as my daughter Jose and they soon became good pals, visiting each other in Belgium and England.

There's no doubt multi-national companies were helping to create a new Europe. In spite of occasional heated arguments at meetings this huge company was a microcosm of the Europe of the future, a Europe without boundaries.

With senior managers of STC/ITT. 1982.

The meetings did have their humorous moments. Once, on route to West Germany from London, we decided that, just for the hell of it, we would oppose whatever the Germans were proposing about one of the new systems. We wanted to see how far we could push them. All meetings were conducted in English. The meeting was chaired by a Herr Bierbaum who was growing more frustrated by the minute at every little opposition we were making. Eventually he exploded, shouting 'You English and your Queen!' To which one of my colleagues replied with an air of quiet shock 'Herr Bierbaum, we can assure you that Her Majesty has nothing at all to do with this system.' Everyone just fell about laughing and, the situation now defused, we got down to normal business.

On another occasion someone cracked a joke in passing at the start of a meeting. Now, one of the Belgian reps, Louis Vervecken, had a habit of walking round the table during meetings. It helped him think. So there we were, minutes into the meeting discussing the intricacies of one design or another, when suddenly strolling Louis bursts out laughing. We looked at each other, looked at the design, wondering what hugely comical error we'd made. Then we realised Louis had just got the joke!

The locations of these meetings rotated between countries, so we all travelled Europe regularly and, as ITT owned Sheraton Hotels, we stayed in the best of places with the best of service.

Our microwave systems production, in spite of the new factory at Basildon, was now being transferred to Germany because their production systems were better and they were better at marketing. As other industries were finding, UK-based production costs were high. It was during the 1960s that the decline of the manufacturing industries began in the UK. British industry was poorly organised, premises were often quite (if not literally) Victorian and restrictive practices on the shop floor weren't helping.

In the early 1970s I was sent on assignment to South Africa, a country I had briefly visited when our troopship docked in Durban during the war. My job was to advise on the setting-up of microwave production there. Microwave communications systems, were a strategic issue for every country. So all governments wanted to produce their own rather than be dependent on someone else.

Based in the capital Pretoria I was both enthralled and shocked by South Africa. The first shock was visiting factories where blacks and whites not only had separate facilities but even separate entrances and exits. Then we realised we had to design two separate production lines because the whites would not work on equipment with the blacks. In fact, they wouldn't touch anything a black man had handled.

Obviously, it didn't make economic sense, but that was the only way they would work. Here I was again face-to-face with the idea that some people were 'sub-human'. It was only going to be a short stay but I didn't like this at all. Aware that I was quickly getting a negative impression of South Africa, my hosts decided to show me the countryside, which was wonderful, except that it also included black townships surrounded by barbed wire. An involuntary shudder went through me. It was the Gulag all over again.

'Look how we look after them! We give them houses!' I was told. The occupants were mainly migrant workers from Uganda, Kenya and Mozambique, mostly working in the goldmines. I felt guilty.

'It's very difficult to train them. We even have to teach them how to hold a spade.' Yes, I understood the meaning. These are 'sub-humans'. This was a shock, a real shock and one I, naïvely, hadn't expected. I was also naïve enough to think the whites were a race of dog lovers because every house had at least two. Then I realised they were all big dogs. Guard dogs. I couldn't wait to get out.

By now we had moved from our three-bedroom, semi-detached bungalow (price £2,400 in 1958) to a house in Broomfield, just outside Chelmsford in Essex, and I was back to travelling Europe. I was beginning to discern my professional capabilities, pitching ideas to international colleagues all the time. In order to survive I was learning and adapting constantly, keeping abreast of every technical development relevant to my field. The money was good, not great, but the work was demanding, exciting, challenging and rewarding.

Training for engineering teaches you to be inventive and to think logically. It is a great discipline for the mind. Britain is now producing fewer and fewer engineers which is not only a great pity, it's also sad loss to the economy and to society.

Telecommunications were developing rapidly in the 1970s and I had to keep updating my skills in both electronics systems and production systems. The design and production of semi-automatic and fully automatic machinery was being accelerated by the introduction of computers and computerised technology. When I first joined STC in 1954 they employed 44,000 people. When I left in 1983 this still successful company, as a result of automation, employed only a few thousand people.

Automation meant fewer production jobs but more maintenance workers to keep the machines going. Better machines were producing better quality goods while computers were often doing the menial work engineers had previously done. As assembly workers became equipment maintenance workers, so engineers were abandoning the slide rule for the computer.

Our first car. A Morris Minor. Gravesend.

Homes were becoming more automated. People were demanding higher quality goods. Cars were becoming affordable. Working people were aspiring to what, a few years earlier, only richer people could have afforded. It was the dawn of the new middle classes.

In Britain in the late 50s and 60s there was also a growing desire for home-ownership. In my travels I didn't see this occurring among my European colleagues until the 70s. I suspect the reason for this was war. For hundreds of years wars had been fought across the continent (but not in Britain) and houses had been regularly damaged or destroyed with little or no chance of compensation. It therefore took longer for the confidence to buy to take hold on mainland Europe.

The microwave industry began to disappear from the UK in the mid to late 70s. Our company's production facilities were closing down. It was time for me to re-think, to adapt once again. I was transferred to a job on the staff of one of our technical directors Jock Marsh who, it turned out, had gone to the same school as my wife Janet – Bell Baxter High in Cupar, Fife.

'Joe, I think we're somewhat related' he said when we met.

'Yes, but only distantly!' I replied.

My job would now involve me in technical planning for the future. Here my mentor was Hector Prior, some eight years older than me and holding the senior position of chief engineer of landline transmission, he certainly transmitted a lot of knowledge to me. I like to think that we made a formidable team. He enabled me to overlay his mathematical knowledge onto some of the planning concepts I had. It was a very fruitful and effective partnership.

I was beginning to see that the future of design lay with computers and I had to introduce the most effective computer-aided support to design and manufacturing. A new and challenging field opened in front of me and I was aware that if I could respond effectively I would be at the cutting edge of this technology within the company. This would mean new horizons and new travel opportunities. Now I was to be involved in exploring interactive systems.

That you have to respond positively to survive remained my belief. Experience had taught me no other. I knew nothing about computers but I knew they were the future and I had to take this on board and quickly. I was gradually to become a computer systems analyst without realising it.

25
My Parents and my Return to Poland

Christmas Eve 1940, the day I was sentenced to the Gulag, was the last day I ever saw my father. Both my parents lived out the war in Luck. Dad worked in the post office under Soviet rule until the Germans occupied Luck on the 24th of June 1941. He continued in his job (postal services being vital to all governments) under German rule until the Soviets re-entered Luck in 1943. They again retained his services until 1946 when Stalin ordered the cleansing of all Poles from the new 'Ukraine'.

The Poles were ordered over the new border and were settled in the new 'western Poland'; that is to say on German land. You have to remember Stalin, Churchill and Roosevelt literally moved Polish land west. Stalin took large swathes of eastern Poland into the Soviet Union and 'communist' Poland was compensated with German lands west of Poland's traditional borders. As if it were a country on wheels, Poland had been shifted one hundred and fifty miles west. These are the borders which still stand.

My parents settled in Jelenia Gora in what had been Hirschberg in East Silesia. This forced mass-migration was the second huge population movement the Poles had experienced in the War. When Germany invaded in 1939 much of western Poland was incorporated into the Reich and the land given to German settlers (many from the Volga and Wolyn areas) while the surviving Poles were forced east into the Government General area of occupied central Poland. Ethnic cleansing is nothing new.

Dad continued working in the postal service in Jelenia Gora. He died in 1971, but not before I had managed to get in touch by a curious serendipity, my luck holding once again.

In 1947 my mother was hospitalised after breaking her leg. In conversation with the lady in the next bed she told her she had lost her son to the Siberian Gulag. She didn't know for certain but she was pretty sure. The lady replied that she was in contact with relations in England and maybe, just maybe, you never know, I could be alive and living there. Mum asked her to give my name to her relations and to put an advert in the papers in England if they could.

One day I was reading the *Dziennik Polski* (*Polish Daily*), when I saw the advert, when I saw my name! The advert gave a contact address in the UK to which I wrote. Two weeks later I got a letter from my parents. You can imagine the joy on both sides!

Enclosed with the letter were three pictures of me and my school pals which mum had rescued from our house in Luck, which the letter told me had been burned out during the Soviet re-invasion in 1943. Also enclosed was an image of the Holy Mother of Ostrabrama, a revered icon second only to Our Lady of Czestochowa in Polish/Lithuanian piety. I had brought this picture of the Holy Mother back from a school trip to Wilno (Vilnius) in 1937. I have it by my bed to this day.

The letter also told me that Zofia and Ignacy were alive and well and also living in Jelenia Gora near my parents. Not long after my arrest on Christmas Eve 1940, Zofia, Ignacy and their two daughters had been thrown off the estate and had taken shelter in my parents' house. It was easier for them to live in a town like Luck than out in the country where resentment for estate managers among the Ukrainian peasantry was positively encouraged by the communist authorities. Nevertheless, Ignacy was soon arrested and deported to Siberia. Because of his age and the injuries he received in the Gulag he was not fit to join the Polish forces after June 1941. He had to wait until 1945 before he could return to Poland. He always provided hope to my family, telling them he was convinced I was still alive. Both he and Zofia were true friends to my parents throughout their lives.

Although I was never to see dad again, I kept in touch by letter, being very careful what was said since all letters were read by the communist authorities. I sent my parents money too as post-war Stalinist Poland was desperately poor. Mum came to visit me in England in 1957, a long journey by train. She was struck by London as a huge, affluent city where she got her first pair of glasses and a new set of teeth! Dad was not allowed to come, of course. He was, basically, kept hostage to ensure mum would return.

With my wife Janet and nineteen-year-old daughter Jose I went to visit mum, in Jelenia Gora in 1971, two months after dad died. I was very apprehensive about going, wondering what I would meet there. Going back to visit on what had been German territory before the War, and was never Polish, was to visit a strange country. I met many friendly, hospitable people but never felt I was in Poland itself – which, of course, I wasn't.

Crossing the Iron Curtain was traumatic for me. This was my first experience of a communist state since leaving Soviet territory in 1943. At the East German border we had to drive into a barbed-wire compound. From the word go, from the first 'Raus!', we were treated as the enemy. As our visas and passports disappeared through a window into a long barracks the guards were hauling everything from the car. As they searched under and into everything memories of the NKVD came flooding back. The guards were behaving extremely rudely, as I had expected they would. My wife and daughter were in total shock.

Meanwhile, in the barracks our documents were undoubtedly being passed from room to room to be inspected by the various *Stasi* officers. I had then to pay for a 'transit visa' to cross the GDR; that is, to pay for every kilometre of the roads on which they were graciously allowing us to travel. We were given our route and instructed not to deviate from it whatsoever.

All of this was making me very apprehensive about what might happen when we got to Poland. Would we be subject to the attentions of the UB (Urzad Bezpieczenstwa), the Polish Security Office?

When the guards had finished their search we had to reload all our luggage ourselves. As we drew out the compound and crossed into the GDR proper, a car pulled out behind us and followed us for about an hour. It was making no attempt to hide (it hardly could as there were very few cars on the road). When it pulled off another appeared on our tail for a while.

The *autobahns* were in a terrible state and not very well signposted. At one point I took a wrong turning and we could see people working in the fields waving and shouting at us to turn back. They obviously didn't see many cars and very few western ones. Our Triumph 13/60 must have stood out like a sore thumb. After this thirty-minute detour, and thanks to the warnings, we got back on the *autobahn* and kept on to the Polish border where the German guards treated us with the same suspicion and disdain.

Crossing into what was now Poland was emotionally difficult for me but the Polish guards were at least polite and they seemed friendly, perhaps partly because I spoke Polish. They also told us the best roads to take, which was encouraging.

I had not set foot on Polish territory since I'd left Luck and it was with utter shock that I entered Kestrzyn, the first town over the border. The adult population was badly dressed and everyone seemed depressed, though the kids crowded excitedly round our western car. Poverty was rife and the 'restaurant' we entered had hardly any food. What a contrast to the old, vibrant, colourful and plentiful Poland I'd left some thirty years before.

At our next stop it was more of the same. The service in the 'restaurant' – which maybe only had one item on the menu – was truly appalling. Not only that, but the waiters were rushing around to serve a coach party from East Germany. I remember I got very angry at this and had what my daughter would call ‘a typical dad moment’, exclaiming ‘But these are the people who crucified Poland during the War!’

When we emerged from the café the AA badge had been pinched from our car. We could see the car from the restaurant window but there was always a crowd round it; so we never saw the culprit. Some youngster had gained a souvenir from the West!

It was an emotional meeting with mum in Jelenia Gora. There was a lot of tearful hugging and kissing. I’ll always remember mum saying ‘It’s a miracle I can still hold you in my arms after all that has happened.’

Pani Zofia and her two daughters, Emilia and Ana, were living nearby. Emilia was four years younger than me and I think it had always been a plan between Mum and Zofia that Emilia and I would marry. Emilia was now married to Vitek, a soldier who’d fought with General Maczek (who lived out his life in Edinburgh) but who’d returned to Poland after the war. Though Vitek still regretted having returned to communist Poland, they were happy together and had two children. A smallholding provided enough to live on and a little left over to sell privately. Things were not so good for Zofia who did not receive a state pension, perhaps because Ignacy had not been a state employee or perhaps because of his time in the Gulag as an enemy of the state.

Mum was living in a fairly nice villa but with four or five families all sharing one kitchen and one toilet. It was a recipe for disaster. You could immediately feel all the domestic stresses and strains of forced, communal living. You even had to take your own loo paper into the toilet. There was no bath or shower, which was particularly difficult for the women who had more or less to barricade themselves in the kitchen for their private ablutions.

The villas themselves, once beautiful, were in a bad state of repair because of shortages of materials. What was available was of poor quality. If you managed to get paint, for example, it was likely to begin to peel within a year. The villas were also fast becoming slums because the people didn’t own them and so didn’t want to put a great deal of effort into them. ‘It’s not mine, so why should I care?’ was an attitude already embedded in the culture. It was one that would only begin to be eroded with the fall of communism.

The people were still living in post-war conditions in what was still a strange land: most being eastern Poles re-settled on what had been German soil until the end of the war. I kept wondering what had happened to the German people who’d lived here. Mum and her compatriots, although they were pleased to be alive, still longed for Luck. There was sadness in our meeting because Dad was dead and so was Ignacy, Zofia's husband, whom I'd wanted to thank for supporting Mum.

These people were struggling to survive. Any problems Poles had had settling in Britain were few compared to the problems of these Polish settlers. They still felt strangers in a strange land, maybe their children less so.

In Jelenia Gora we were watched and followed by police all the time. On arrival we had to register at the local police station where we were asked to surrender our passports. Janet, being a somewhat feisty Scots woman, refused. She was told this was the law.

'I don't care what your law says. You can arrest me if you like, but we are not surrendering our passports.' She was adamant about this and they backed down. She was rightly afraid they might not let me out again.

The paranoia of officialdom, not of the people, was apparent everywhere. An absurd example involved a local crystal glass factory which I wanted to visit out of interest. The factory director refused me permission on the grounds that I might be a spy. I'm sure what information I could have gleaned from such a visit would have been as useful to western military intelligence as a crystal ball! But, then, he'd probably have lost his position for letting me visit.

Officially, I was the 'enemy', but socially I was a friend who'd come back. As many of the resettled population were from in and around Luck a lot of folk wanted to meet up. It would have taken two months to fulfil such obligations and we only had two weeks.

We were always drawing curious crowds. Everyone wanted to know about life in the West and how the Poles were getting on in Britain. This was mingled with a degree of resentment towards Britain, which they felt had abandoned them. Nevertheless, when it was time to leave you could see envy in their eyes. We were going to the free West, the ambition of every Pole.

The families we met in 1971 had little food. There had been some post-war austerity in Britain in the late 40s/early 50s, but this was something else. To begin with there was very little meat except chicken, people keeping their own chickens for eggs. Most other meat was 'exported' to the USSR or allocated to Russian soldiers in Poland. Being able to keep a few chickens for your family was one bonus of the communist government's decision not to 'collectivise' Polish farms. Moscow might have wanted this but Warsaw realised the Poles would resist all the way. To achieve this they would have had to impose a Ukrainian style famine, which they were not prepared to do. Farmers were still told what they had to produce but naturally held on to what they could for bartering locally. Chickens were valuable and not often eaten except on special days. As we were special guests we were served chicken in almost every home we visited - a great privilege.

If you knew a local farmer you might also get a piece of a pig for money or in exchange for another commodity. Without the black market (which, in reality, was simply a very restricted free market) people would just not have survived.

There was plenty of vodka. The State made sure of that. It was *the* social anaesthetic and people drank a lot. During our visit the vodka was flowing and Janet got drunk for the first time in her life. She was practically teetotal but couldn't refuse all the toasts – to friendship, to Poland, to the old days, to Scotland and so forth.

Only two types of business, taxis and hairdressers, appeared to be permitted. Both were cheap and plentiful. Other shops were state controlled, even the local grocers.

One day, spotting a hosiery shop, Janet went in to buy some tights. She wanted size five. They only had size three. The assistant told her she'd get the size she wanted in a town some ten miles away. In the 'retail sector' of this planned economy there was no catering for any real market. The only things you could always get were vinegar, vodka and matches. My half-brother Jan's three boys came rushing in full of excitement one morning shouting 'There's going to be bananas!' They were aged between fourteen and twenty and they had never seen or tasted a banana. We went and queued, as you always had to do, and managed to get some of the few bananas left. I then had to show them how to peel a banana.

As we had hard currency we could use the PEWEX shops which were well-stocked with food, clothes and toys from the West as well as locally manufactured goods being sold to help with the 'balance of payments'. In these shops you always had to watch your wallet as they were often visited by thieves on the look out for richer pickings or even some hard currency.

Our car, though, was safe enough from thieves. It stood out like a sore thumb. Anyone stealing it would have been caught immediately. They might even have been seen stealing it since we were followed by UB officers on foot or in a car most of the time. They were quite visible, wanting us to know we were being watched.

There were a few German tourists in the area: East Germans who had lived there before the war or their children. Close to both the East German and the Czech borders, it was a beautiful region and a traditional spa area where people came to take the waters. On one excursion we took a chairlift up the mountains from where we could see into what is now the Czech Republic. The mountaintop restaurant was busy with German tourists. As we were speaking Polish we were being ignored – even GDR currency was worth more than the zloty in those days. I had to inform the waiters that we were foreign tourists as well before they paid us any attention.

The visit was a mixture of happiness and sadness. It was great to meet old friends and good to be able to visit dad's grave. But it was sad that I was meeting all my old friends six hundred miles west of Luck in an area which was not Poland. It was good that mum was fairly secure and that she was enjoying looking after her grandchildren. But wages were meagre and both parents often had to work to survive. I'd often sent, and would continue to send, money and parcels of clothes. When we were there a lot of parcels were flowing in from the USA where ten million Poles now live.

Mum was sad that she couldn't show me Luck but there was no point in my trying to take her into the Ukraine. I simply wouldn't have got the visa.

Although we would make one more visit two years later, the truth was I just couldn't tolerate the political and economic climate there My wife and daughter, while they were taken with the friendship and hospitality they met with, were appalled at the lives people had to live. I felt uncomfortable, even ashamed, at what I was exposing them to. I really didn't want to put them through it again.

When it was time to go and we set off I wanted to get back to the West as quickly as possible. The Polish border guards were courteous but as soon as we crossed the Oder we were followed by military jeeps with either East German or Soviet soldiers in them. We were virtually escorted to the West German border.

In the compound, aggressive guards stripped our car. They didn't bother looking in our luggage. They weren't interested. They were looking only for defectors from their communist paradise.

The journey back was sad and almost silent. There was so much to think about. I had travelled all over Europe and it struck me that Poland was the only country still suffering because of the war that had ended thirty years before. It had suffered physical damage and lost too much of its population. Not only that, but at Katyn its most creative people, its intellectuals and future leaders had been massacred.

I sensed the Poles were willing to forgive the Germans, but never the Russians. I got this from everyone I met, family and friends. It was not only because they had murdered in such an underhand way, but also because they wouldn't admit to it.

Eventually, of course, Yeltsin admitted it. But he never apologised. Willie Brandt went to Warsaw to apologise for what Germany had done there. But not the Russians. Until they do, Poland will never be friendly towards them.

There was sadness everywhere in Poland. And I could detect

unrest. An uprising was only a matter of time. Small uprisings had occurred in the 50s in Gdansk, Stettin and Poznan and had been repeated. They'd not been directed by the intelligentsia, but by ordinary workers. I could sense a time-bomb ticking and I was afraid of another Polish bloodbath. The election of Pope John Paul II in 1978 changed everything. After that received letters full of hope. John Paul had, and still has, tremendous standing in Poland. He's still treated as a saint, not so much because of his piety but because he brought tremendous hope and then change.

Travelling into West Germany we entered a dynamic country so different from the communist east. We were met with a riot of colour and light and the sound of laughter. We went straight into a restaurant where Janet had a huge steak. After two weeks of chicken I don't think she ever enjoyed a steak as much as that one.

At the Dutch border I asked the tourist information lady to book us a hotel at Arnhem. Without discussing any details, she booked us one situated only a hundred yards from the cemetery where many of my colleagues were buried. I think, instinctively and through experience, she knew exactly where we were going - and why. After booking into the hotel we went down for dinner. On our table were a small Dutch and small British flag, a symbol of Dutch appreciation for the role we had played in their liberation nearly three decades before.

Visiting Arnhem to pay homage to my colleagues and to show Janet and Jose where we'd fought brought a sadness too. What

The memorial at Arnhem to those who fought in the battle.

had it all been for? What had Poland's reward been?

In the summer of 1973 we did visit again, this time to the Baltic city of Szczecin where my cousin Josef Glogolski was living. He was a cousin on my mother's side and had been brought up in Luck. He arranged for mum to be there at the same time.

Josef had been a reserve officer in the Polish Army. He was captured by the Germans and imprisoned in an 'Offlag', an Officers' Camp, in eastern Germany.

I also met a school friend in Szczecin, Przemek Halacinski, whose father had been murdered at Katyn. His mother had lost her mind as a result of this and my mum had helped him a lot through the war. He was now a Professor of Economics in Warsaw, but no Marxist.

We also met a woman, slightly younger than me who, as a child, had been in a German concentration camp with her mother. She told us how they had been lined up against a wall by the guards who had marked their hands with a cross – this would be a 'selection' – marking them for execution. But she had a lot of thick hair (it must have been soon after their arrival in the camp) and had just had a dousing in water and disinfectant. Her mother was thus able to use her thick hair to erase the cross from their hands. So, they survived.

On a lighter note, I also remember that my daughter Jose had with her a Pink Floyd album, *Dark Side of the Moon*, which she gave to some teenage boys. Needless to say, they were more than delighted.

Conditions were slightly easier on this visit because recent troubles in Posnan and Szczecin had forced the Brezhnev government in Moscow to back-pedal a bit. But it was still very oppressive: food shortages continued and the secret police were still very much in evidence. My cousin showed us pictures he had taken of the demonstrations in Szczecin and of their brutal suppression.

The country was still full of badly dressed people, many living in those brutalist communist blocks, and there were still few smiling faces. Everywhere we went we saw slogan after slogan praising the regime and 'Fraternal Brotherhood' with the USSR.

In spite of all this we enjoyed the family visit – though there was a shadow. Nothing was said, but mum, I think, knew this would be our last meeting. On the previous visit she had given me dad's wedding ring which had her name (Katarzyna) inscribed on it. This, I had given to Jose as an heirloom, a memento of her Polish roots. Now mum wanted to give me her wedding ring – which I couldn't accept. It was a tearful moment. She had nothing else to give.

Mum died in 1981.

26
A German Sojourn

From the early 70s onwards I began to detect a loss of influence and of markets for the British arm of ITT. The British were finding it difficult to adapt to new conditions. It was very noticeable that although our electronic engineering concepts were good our implementation of them was poor and our marketing skills not much better. Perhaps management was still stuck in the old imperial dream but in reality Commonwealth countries felt no real affinity to Britain when it came to business, to hard cash.

Britain's role in Europe was diminished. ITT had companies or branches in every European country and these continental firms all co-operated with each other while the British arm was still trying to be different and thinking it could still lead. It couldn't. It was sad to see British influence in ITT diminishing while the Germans, French and Belgians were racing ahead.

Computers were being used for financial systems by many companies in the post-war years but it was not until the advent of transistors and micro-processors in the mid-60s that they really took off in terms of engineering and design. What we were really looking for as a company were systems which could cope with graphics. Dependence on man-made drawings, laborious and labour intensive, had always been our Achilles' heel. If only we could get computers to cope with this!

As computers were beginning to be used in production planning my task was to see what was available in the market and to recommend systems for design and engineering applications. This meant more travel to meet with all the big computer manufacturers. Most systems, of course, came from America, many developed at MIT, IBM being the leader. But it was difficult to establish a consensus between ITT in Europe and its British counterpart STC. While most of the European companies were going for the IBM system for financial and production planning STC opted for Honeywell; and while Europe went with Computer Vision for design engineering STC decided to adopt Calma. All the time STC was being different. It was affecting profits and diminishing any chance of a leading role.

In our field and, I think, in general business terms, Britain was on the margin rather than at the centre and was constantly rejecting European ideas and systems. Britain was, in fact, struggling to find its place in the new Europe. It was failing to adapt while countries like France and Germany were doing so eagerly. This was partly a legacy of having been a colonial power. But then France had also had an empire. There was something in the mentality of British people holding them back. Having been a world power Britain was now being asked to play a secondary role - and was finding this unacceptable. They couldn't compromise with Europe.

The transfer of information is the lifeblood of any company, especially a telecommunications company, and my job was to find the optimum system to serve us all. No small or single company could by then develop huge systems on its own. Multi-national solutions had become the only viable route. By being 'different' our British set-up was damning itself. Instead of an easy flow of information across the multi-national everything had to be (re)-interpreted through different systems for STC.

On the macro-economic scale Germany and France had now become more important to the USA than Britain and it was significant to me that in both these countries management was rapidly embracing the English language while American dollars were pouring in to offset any threat from communism. Rich people don't vote communist.

As I was becoming frustrated in my role on the technical directorate staff I also knew that various projects were being set up in West Germany. One in particular caught my attention. In Stuttgart the company was planning to develop semi-automated and then fully-automated production lines to make telecom equipment. This was being done by an international team led by the Germans.

Jose's graduation.
Leeds University. June 1973.

My daughter Jose, who had left for University in 1970, graduated in 1973 then completed a post-graduate degree in 1975, was now married and settled. So Janet and I discussed a move to Germany, the post being for a minimum of three years. I had reservations. We both had. How would I feel about living and working in Germany on a more permanent basis? Would I be able to integrate? Would they accept me? How would Janet take to it?

We saw it as a challenge and we took the plunge in January 1976. I had visited Stuttgart a year before. Severely bombed in the war, it was now almost entirely newly-built. It was also a powerhouse of industrial renewal. A city of close to one million people, it was home to Mercedes, Bosch and Porsche and extremely prosperous.

Moving there we could see prosperity growing at a rate well beyond anything we had seen in Britain: new shops were springing up alongside top-class hotels and restaurants; people were well-fed and well-clothed. American capital was flowing in. This was the bastion against communism.

Everything was clean in contrast to the grime of London and people were socially responsible citizens in ways the British were yet to learn (and are still learning). They were already recycling bottles and paper. I saw my first 'wheelie bins'. There was a social discipline in schools, in factories, in everyday living. West Germans were strides ahead. This became more apparent every time I returned to London (at three-monthly intervals) to make my reports.

The booming economy had brought in millions of workers from Turkey, Greece, Yugoslavia, Portugal, Italy and Spain to cope with the expansion. Under the Nazis the German economy had also depended on foreign workers – slaves – but these were *gast-arbeiter*, 'guest workers'.

Few of these workers arrived with any German, so management had a problem: how to communicate. The answer they came up with was devastatingly simple: pictorial instructions for the factory floor. This was meticulously planned and executed and it paid dividends. In every factory everything was clearly marked and defined. Meanwhile, on my visits back to London, management were taking no such measures for new Commonwealth workers who had little or no English. I remember discussing this with one of my bosses. His attitude was that all instructions are and will be in English. What if they don't speak English? They'll have to learn!

Which world was going to win was obvious. We were so far behind. Our continental factories were also spotlessly clean in contrast to our shop floor in Southgate. I remember noticing dropped components on the shop floor there, for example, and pointing out to the person that this was money lying on the floor.

'Why don't you pick it up?' I asked.

'It's not my job', she said.

You just didn't get that attitude in Europe, especially in Germany. I don't think it's a 'worker's attitude'. It stems from management who determine the culture of the workplace. There was a lack of leadership in all spheres for which Britain is still paying. I was certainly more attuned to the continental way of thinking and doing. But then I was – I am – a continental man.

In Stuttgart our flat was almost next door to the factory. Not only did this mean no commuting problems; it also meant I could have lunch with Janet every day, which meant a lot to us especially in those first days in a new country.

I had very little German to start with, and Janet had none, but we were welcomed by our German colleagues with open arms. Janet was made to feel at home which was important as I was still travelling every week in Germany and beyond on company business. At weekends groups of us would go on walking trips – *wandertag* – with German and other European colleagues and their wives. The social bonding meant that Janet could go on these trips even if I was away. We were an international community.

Everyone spoke English to a greater or lesser degree but I wanted to learn German effectively. I was determined from the start that at the end of my three year sojourn I would make my farewell speech in German, which I did. In my spare time I watched German TV with a 'Teach Yourself German' book in front of me and eventually became proficient.

I had also brought over my car, now a Triumph Toledo, which, naturally, had the driver's seat on the 'wrong' side. This started a little joke we used to play at traffic lights. Big German cars would impatiently toot at you if you didn't move immediately the lights changed. I would pull away slowly with Janet, in what looked like the driver's seat, looking backwards at the impatient motorists who had a look of horror on their faces as they thought she was driving off looking behind her!

As on my previous visits to West Germany, my Polish origins were never mentioned. I was treated as an Englishman, but Janet would always underline that she was Scottish, not English, and often broke into broad Scots if she wanted to communicate something private in company. Although many Scots words and pronunciations are more German-sounding than southern (or 'standard') English, our German colleagues could never follow what she was saying at these moments.

One of my 'German' colleagues, Manfred Salwitzek, was of Polish descent. He was originally from Silesia. Knowing my background, he would often speak to me in Polish when we were alone. He was the only one to do that.

Our team leader, Jorgen Duhm, was from Danzig/Gdansk. He had a huge moustache. 'Put a pointy Prussian helmet on him', said Janet, 'and he'd be the spit of Kaiser Willie.'

We all visited each other, had get-togethers and parties. Wine was plentiful in the area, especially good quality red wines that you don't often get in Britain, which seems to import mostly white wines from Germany. We once had an outing to a wine-tasting event with one hundred and forty local wines to choose from. We never got past the first twelve. We were drunk.

I also got to know 'ice wine' which the Germans choose for big celebrations where we might have champagne. Made from grapes picked after the first frost it is a delicious non-sparkling wine. It was in Germany I began to learn about and to appreciate wine.

While my work was fulfilling and our social life was good, it was impossible to escape German bureaucracy. Wherever you lived you had to register with the police; if you moved you had to de-register and then re-register your new abode whether you were foreign or a native German. It was impossible for the police not to know where you were. British people would probably be horrified at this, but it is, I suppose, a question of balancing individual freedom with protecting the freedom and safety of the majority.

My car was to prove a potential problem. A Spanish colleague had for three years tried unsuccessfully to get his (Spanish) car past the German equivalent of an MOT. I think they really wanted you to buy a German car. The MOT was very strict and it already had emission limits for the exhaust thirty years before the UK introduced such environmental measures. At the first attempt my car failed because of the level of emissions, because the exhaust was two inches too long(!) and because the headlights needed realigned. When I got these fixed and took the car back for its test again it was a different mechanic. He'd never had any experience of a British car before and was quite excited by it. So I told him to take it for a drive if he wanted. He jumped at the offer and enjoyed himself thoroughly. I got my clearance, I think, on the back of that!

In leaving Germany three years later I had to go through the same procedure again to de-register my car in Germany. I was given a certificate to say my car was roadworthy to return to Britain with. As if anyone in Britain would have cared!

This was all part of the German culture of social responsibility. Living in a flat meant you had your *ker woche*, your 'duty week', when you had to clean the stair and the pavement outside the block of flats right up to the kerbside. I never heard of any arguments among neighbours over this sort of thing. People accepted such duties without demur.

At weekends when we went touring, the *autobahns* were clear of heavy lorries from twelve midday on a Saturday (when the shops closed) until Monday morning. The roads were freed up for family and pleasure trips, and the whole of Germany seemed to take advantage. We made trips to places like Munich and Nuremberg and through the Black Forest.

I spent three happy and fulfilling years in Germany and still keep in touch with international colleagues from that time. As my sojourn came to an end I had three options facing me. I was offered a position with ITT at Forth Worth, Illinois, in the USA. I could stay in Germany or I could return to Britain.

Around this time Janet began to feel unwell. We were afraid it might be cancer. After many tests it would eventually be diagnosed as Crohn's Disease, but at this time we just weren't sure. Deep down I knew she also wanted to get back to Britain to be near her daughters. She had got used to me working close to home and the only stipulation she made was that whatever job I took it would not involve a long daily commute. So Britain it was. But before I could leave I was asked to find my replacement for the Stuttgart job, which I did. The man I found, John Blinkhorne, was most effective and the post was to prove the launch-pad for his career. John and his wife Jenny were the most remarkable people. They adopted three children from the same family and brought them up.

27
St Andrews

Back in Britain I took up a position on the staff of STC's operations directorate exploring the possibilities of computer integrated planning and manufacture. It was a difficult task as STC was still out on a limb, not using IBM systems. It was going through a difficult period while its sister companies on the continent were becoming more powerful. The STC dilemma was: either co-operate with those sister companies or co-operate with UK companies in developing new microprocessor and transistor systems.

The Post Office was resistant to having an American company involved in developing communications systems they would operate as British Telecom. STC was falling between two stools. The result was that ITT agreed to an STC management buy-out in 1980-81. In effect, they sold it off. So I was now to work for a junior partner in co-operation with bigger companies like GEC and Plessey. The GPO/Telecom contracts were huge and at a Downing Street meeting chaired by Margaret Thatcher it was decided three companies were too many. The contracts went to the two big boys and STC lost out. It was now against the wall.

British Telecom also decided to open contract bidding to foreign companies like Erikson and Siemens. UK companies were poor followers because of their inability to innovate and adapt at a time when Britain was losing its manufacturing base. I was heading for retirement knowing the battle with Plessey and GEC would be lost. Generally it was a depressing scenario though I was still travelling a bit in Europe, and occasionally to the USA, to look at emerging technologies.

Visiting New Orleans and Philadelphia I was struck by the co-operation between the companies developing President Reagan's 'Star Wars' programme and between those companies and the American universities. The US Government was demanding a total commitment from all companies and universities to offset what it saw as the Soviet threat. The US military was very much in the driving seat.

During this period Janet was suffering more, though we still didn't know what the illness was. Some of the doctors thought it was cancer. We were concerned. Janet had by now also had enough of life in London and we decided she should go back to Fife in advance of my actual retirement. She went to stay with her favourite niece, Margaret, in St Andrews.

Margaret immediately took her to see her GP, who referred her straight to Ninewells, the teaching hospital in Dundee. After a battery of tests, they too suspected cancer. I went up to be at her side. She was losing strength rapidly. She was eating little because it was causing pain and discomfort. The consultant surgeon now thought it wasn't cancer but they needed to operate to know for sure. We had to build up her strength for the operation, which was to find damage to the small intestine but no cancer.

I told my boss that my responsibility was to be with my wife and that I wanted to retire as soon as it could be managed. He agreed, but said to me 'I'll tell you one thing, Joe. You'll never retire!' He was to be proved right.

Selling the London house, we bought a bungalow in St Andrews and, at the age of sixty-three, I retired officially with a big party for all my European friends. I was deeply touched and immensely pleased that so many flew over for the celebration.

Janet had made a good recovery but the pain was never far away. In 1986 she took another downward turn and this time, when they operated, they discovered she had Crohn's Disease, which is incurable, though the symptoms can be relieved by painkillers, steroids and diet. She was put on steroids. At least she was close to her family in nearby Cupar.

I was helping to look after her and also playing golf (how could you live in St Andrews and not?) to keep fit and active. Golf was an escape from worrying about Janet. She was in constant pain but life was tolerable. She had a strong will and character. But she must have thought I was dwindling myself or not doing enough to keep busy, so she encouraged me to find something else.

The result was that I took a part-time lecturing post in computers for business support at the Dundee Institute of Technology, soon to become the University of Abertay. My professional life had always involved giving presentations both to management and workforce, so I took to teaching like the proverbial duck to water and was enjoying it immensely.

Janet deteriorated again and was referred back to Ninewells for the third time in early 1988. Her situation was now becoming critical. They tried various treatments before deciding to operate for one last time that December. She was admitted to Kings Cross Hospital, also in Dundee.

The operation had been scheduled for after Christmas but it suddenly had to be brought forward. Our daughter Jose was stranded in Germany and couldn't be at her mother's side. She was distraught. The consultant said there was no reason for her to return, but nonetheless she tried to do the impossible – to get a flight out of Munich on Christmas Eve. She had come up from England to visit her mum before her operation and had been told by the consultant there was no reason for her to alter her travelling plans. Even when the operation was suddenly brought forward we did not know until the last day that Janet's chances of survival were slim.

Quite fortuitously and for a reason she still cannot explain, Janice (Janet's daughter from her first marriage) had a premonition that she had to get home for Christmas. It was very difficult for her to get an exit visa from Saudi Arabia, where she was working, but she managed to use a colleague's and came home quite unexpectedly. So, as it turned out, she was at her mother's side during her last hours.

When she was unconscious after the operation on December 24th, it was apparent that Janet was dying. I tried to find a Church of Scotland minister for her. But, being Christmas Eve, this wasn't possible. However, the hospital did locate the Roman Catholic chaplain. I asked him if he would give Janet the Last Rites. At first, he refused because she was not Roman Catholic. Under canon law he could not administer the Last Rites to a non-Catholic.

'Father', I asked him. 'Do you honestly think that God distinguishes between Catholics and Protestants?'

This shook him, I think, and he agreed to my request. He administered the Last Rites and gave her Holy Communion. Janet died shortly afterwards at 6pm.

Her funeral took place on St Stephen's Day (Boxing Day) 1988 at Ceres Church where she had been baptised. She was buried with all her family and friends there. A close friend of the family, a piper, played a lament as her coffin was lowered into the earth.

28
Polska Restituta

I went through a difficult period of adjustment after Janet died. I couldn't stand the silence at home, a home that had always been full of noise and life. I joined every club there was. I was golfing, playing bowls and supporting the local football team St Andrew's United. I had to escape the silence. I had to be with people. I was undergoing another process of change. What I was looking for was something really challenging, only I didn't realise it.

I had also been keeping an eye on what was happening in Poland, watching with hope, and some trepidation, the rise of *Solidarność*; and then greeting the new *Solidarność* Government in 1989 with even more hope, and some relief.

In 1991, when I was out golfing with a friend Bolek ('Bill') Szymanski, my life was to change again. Bill and his wife had been teachers in Wales but had retired to St Andrew's, where his wife was from. Bill knew I'd had difficulty coming to terms with Janet's death. That day, on the golf course, he showed me an advert he'd cut from a newspaper. A charity called BESO, the British Executive Service Overseas, were looking for people with business experience to help Polish companies make the move from the old command economy to a new market economy.

'This is for you, Joe – right up your street', he said.

I was intrigued. I wrote immediately to BESO and received a reply from a Mr Jan Kossakowski saying that as I was bi-lingual and given my business and lecturing experience I was exactly the kind of person they were looking for. I was bi-lingual, but my Polish was not up-to-date, especially in technical vocabulary. But fate provided a solution. There were several Polish post-graduate students at the University of St Andrew's. One of them, Dr Ana Lisowska, helped me update my technical vocabulary and motivated me greatly to accept the BESO challenge. I had helped Ana, who was divorced with a six year old son, to settle in the town. She was only too glad to help me as she was a very patriotic Pole.

My first BESO assignment, which came almost immediately, was to advise the Warsaw Pump and Fittings Company. This was to be my first experience of how companies were operating, or trying to operate, in the new Poland.

This was my chance to do something for a 'free' Poland. But

how 'free' was it? Politically it was free but, as I was to find out very quickly, in industry and the civil service, at both local and national levels, the thinking was still definitely communist. How could it have been otherwise? The transition was going to be difficult.

I soon realised that talking to forty- and fifty-year-olds about new business structures, new ideas, was largely a waste of time. They wouldn't be able to make that transition. They had no initiative. One of the managers at the Warsaw pump factory said to me, 'Look, Joe, a few years ago if you showed any initiative here you were simply knocked back. All planning came from above. You'll have to give us time to change.' How much time? He reckoned it would take two generations to eradicate what he called 'the communist disease we are suffering from mentally'.

The task before them was huge, almost impossible. All their old 'markets' in the Soviet Union, Bulgaria etc had collapsed virtually overnight along with communism, but they were not yet ready to compete in the 'West', in the markets of the free world. They were going to have to go through a hard period when most of the established 'firms' would collapse. There was no point in throwing money at them. They were outmoded, outdated. This money would simply have been misused. Furthermore, the infrastructure was very dilapidated. In the early 1990s the management systems were simply not in place for Poland to profit from its new freedom.

I decided to concentrate my efforts on the younger generation, those in their twenties and thirties, and to run courses on project and financial management. Such things had never been done at company level before. This had been the preserve of the government civil service in communist Warsaw. Being Polish, and speaking Polish, I knew how they thought as Poles. I also had experienced what communism could do to people physically and mentally. The older people were resistant to change, not simply because they had been 'communised' but because they were afraid to give up the reins of power, afraid for their meagre pensions. And understandably so.

I made four separate three-week visits to the Warsaw Pump and Fittings Company. I knew my first visit had been a shock to them as they realised just how far behind they were. On each visit I met other companies seeking help and advice and, between my assignments to the pump factory to follow up on the implementation of ideas, I visited these other businesses too.

Even on that first assignment in Warsaw I met people I knew from Luck. These included old friends from school such as Nina Kepska, who had been my sweetheart for a time and whose mum had fed me nourishing goat's milk in the early days of the war.

These people were now in various walks of life, from writers and theatre directors to army generals.

What struck me, especially in those early visits, was how we had all grown apart. I don't mean that we'd simply grown apart as people will after fifty years; but there were deep differences of political, social and economic experience. Many of my old schoolmates had been 'communised' to a greater or lesser degree. As a result there was a largely unspoken agreement that we should just enjoy each other's company and avoid the touchy subject of politics.

An old school friend Zdzisiek Wincentak had become a General in the Polish Air Force and had known General Berling. He was a convinced communist as he said he had been at school (though I certainly never knew that at the time). He didn't see all the cruelties communism had caused the people. Like many of the old communists I met he turned a blind eye or excused it.

Time and again it was underlined to me by other old friends and new acquaintances that they had been abandoned by Britain and the USA. This was, of course, only the truth. It made me think that, while I had had three or four difficult years, they had had half a century of difficulty and that that had had a big effect on their thinking and their attitudes, which had also been distorted by propaganda. They badly needed to experience democracy. Despite having suffered grievously they had to get away from simply blaming others for their misfortunes.

In Warsaw there were still many bullet holes in the buildings. That, I'd expected. What I hadn't expected were the bullet holes in people's minds and souls.

To begin with, I was welcomed as the guy from the West who knew it all and who had all the money. In the early 90s I must have seemed like a millionaire to many Poles. Indeed, I almost was in their economy; an economy which paid teachers and doctors the equivalent of £20 or £30 per month by our standards.

There was no point in saying to businesses 'do as we do in the West'. You had to explain from first principles, explain the logic, share an understanding before people could move forward on their own two feet. In order to do this, you had to adopt some of their thinking, put yourself in their shoes.

There was, of course, almost universal praise for the Iron Lady, Margaret Thatcher, in those first days of freedom and it was difficult to dispel the myth of the great and good lady. If you spoke of how Thatcherite policies had also laid waste whole industries and communities in the UK you got the distinct feeling they weren't really listening. Not yet. Since then they have learned that capitalism can bite as well as feed.

In the 1990s the Poles still hated the old oppression. They weren't yet able to examine their own thinking to see how it had been corroded. One thing I learned for sure was that communism attacks the brain as well as the heart and soul.

On the other hand, I still met a lot of resentment over Katyn and Russia's steadfast refusal to apologise. That the Soviets had deliberately set out to murder the flower of the nation so coldly is something deeply ingrained in the psyche of every free Pole. Poland has now welcomed NATO anti-ballistic missiles into Poland. They are very welcome. And Russia knows why: Katyn. The Soviet Empire fell apart because of its greed. It swallowed too many countries to control. Now all those former Soviet bloc countries are totally committed to the west and totally against Russian influence.

The role of the Church in post-communist Poland also disturbed me. After Janet's death on Christmas Eve 1988 I had started to go back to church, but my experiences in the new Poland led me to stop going once again. I wouldn't say I lost my faith in God; nor do I reject the basic principles the Church preaches, but I am concerned and disillusioned by the Church in Poland and by modern Polish priests.

On my first BESO visits I became aware of the Church's tremendous power. Undoubtedly, it had been a force for the good in communist Poland, as a mainstay of national identity and traditions. But, as power corrupts, I think the Church became corrupted after the fall of communism. It hasn't yet found its true place in the country.

There's no doubt Pope John Paul II was a great patriot and that the Church, with the aid of Ronald Reagan and the CIA, was crucial in bringing down communism. But then it became intoxicated with its own power. Nowadays, a lot of Polish priests have live-in partners and I've never met a poor priest there. The country is full of brand new churches and new priests' houses. They seem to me to be living in the lap of luxury at the poorer people's expense.

Until the mid 1990s I was still teaching at Abertay. My retirement was proving a busy one. I encouraged some of the professors at Abertay to set up business projects with sister universities in Poland. Three professors in particular, David Kirk, Jim Paterson and Jeremy Schmidt, were very positive. They sourced monies from the UK Know How Fund to visit Poland and eventually to set up a partnership with the Kielce Polytechnic. We brought some professors over from Kielce to Dundee. I remember one commenting: 'You are educating engineers. We are educating PhDs. Your people are practical. Ours are theoretical. We need to change our emphasis to the practical application of engineering.'

Subsequently, Abertay established partnerships and/or joint projects with six polytechnics: Lodz, Gdansk, Wroclaw, Katowice, Slask and Kielce before spreading its wings into Lithuania, Bielorus and Ukraine. The role Abertay played in eastern Europe was significant. It had an undergraduate exchange with Lodz and conferred an honorary degree on Leszek Balcerowcz, the Depute Prime Minister of Poland and the Minister of Finance responsible for the programme which not only curbed inflation in Poland in the 90s but became the model for other post-communist economies in how to tackle inflation. An affable, approachable, but very single-minded and intelligent man, Leszek was to become Governor of the Bank of Poland before retiring.

I also worked in Zielona Gora with sixteen small to medium-sized companies, all trying to develop their potential in this former industrial centre which had become an economic ruin. After working with them for eighteen months, I suggested they send a delegation to Fife to see how the transition from coal to light industry had been managed there. Forty delegates came by bus to meet with development agencies, colleges, universities and businesses. It was a great success and it must have pleased somebody over here. It led to my being honoured, for my efforts to set up business and educational links with Poland, on December 12th 2000 when I received an MBE.

Buckingham Palace, 12 December 2000.
Josef receives his MBE from Her Majesty Queen Elizabeth in recognition of his work in promoting British-Polish industrial and educational links.

How this came about I'm not sure, though I did meet Princess Anne, the Royal Patron of BESO on two occasions. I accepted the honour from Her Majesty the Queen because it was important to me that the kind of work I was doing was recognised and because, apart from military honours in wartime, very few Poles had been honoured in this way. After the war I was officially an 'alien'; now I was recognised by the Crown. It helped in a small way to make up for the British Government's failure to acknowledge the part Poles played in the liberation of Europe.

Do we still await some kind of belated recognition from the British government? An apology perhaps for our exclusion from the 1945 Victory Parade? Both would be too little too late. We were the fourth biggest allied force until the French colonial forces 'rejoined' the war effort during the invasion of North Africa. Excluding the Home Army, the largest underground army in Occupied Europe, we were 500,000 strong and we were the only country which stood by Britain throughout the war, from beginning to end. 'For Your Freedom and Ours'. This was the bond, the unbroken alliance. Britain broke that bond in 1945. Obviously what really mattered was that Poland had been fed to the Stalinist wolves. That hurt is still felt in Poland. But we were also hurt and dishonoured by being wiped from official history in May 1945. The ultimate blame, of course, lies with Russia. Did Britain act shamefully? I don't know. It's not for me to say. My feeling is simply one of sadness: the way you feel when you know you have been let down by someone who was, perhaps, your best friend.

The Dutch people certainly never let us down. In 2004 – the 60th anniversary of the battle of Arnhem – they honoured all the veterans of Operation Market Garden. We were treated like royalty. As we crossed the Bridge, people of all ages, men, women and children, lined the route and threw flowers at our feet as the marching bands played the theme music from the film *A Bridge Too Far*.

On that visit I actually found the spot where we had dug in as the Germans bombarded us. I suddenly remembered how the ground shook so violently that the apples began to drop from nearby trees and the local women, starving, rushed out to collect them in their aprons in spite of all the danger.

The 1st Polish Parachute Brigade was also honoured by Holland in May 2006 when it was presented with the highest Dutch military distinction: the Order of King William of Orange. (In some Scottish circles that might make me unique – a Catholic Orangeman!)

When Queen Wilhelmina returned to Holland from Britain after the war, she wanted to award medals to all the fighting units who had helped liberate her country but the USSR and its puppet government in Warsaw objected. The Americans, for example, received the Order, but we were ignored as we had been by Churchill and the British Government who forbade us from taking part in the Victory Parade in London in 1945 for fear of offending Uncle Joe. Forgotten yes, and insulted too.

Prince Bernhardt, consort of Queen Juliana (Wilhelmina's daughter) never gave up the fight to recognise the Free Polish Forces' contribution to Holland's liberation (they had spent their honeymoon in Poland). But nothing could be done as long as Moscow ruled in Warsaw. It was only after 1989 that their daughter Queen Beatrix was able to promote the cause to good effect.

At the ceremony in the Hague when the Polish Parachute Brigade was honoured, General Sosabowski posthumously received the Order of the Bronze Lion. The British government had not even given him a military pension and he had lived out his life supported by colleagues and friends. At the Hague, surviving colleagues from all around the world came together again, including, to my absolute delight, Leszek Ziolowski, my best friend from primary school and one of the 'three musketeers'.

Queen Beatrix of the Netherlands awards the Royal Order of King William of Orange to the 1st (Polish) Independent Parachute Brigade for its distinguished and outstanding acts of bravery, skill and devotion to duty at the Battle of Arnhem

If London had let us down, our British colleagues didn't. Major Urquhart, who'd tried to warn General Browning against the Arnhem plan, raised money to erect a statue to General Sosabowski in Driel where the bulk of the Polish Arnhem forces had fought. It was unveiled in November 2006.

It was through my BESO work that I met Lech Walesa. I'd been visiting the Doraco Construction Company in Gdansk pretty regularly, advising them on financial and production management. I think I made six visits in three years. Around that time, BESO director Iain McConnell was organising a Burns Supper. He'd held one in New York the previous year with Kofi Anan, Secretary General of the UN, as main speaker. Now he couldn't get Mikhail Gorbachev because of a clash of dates. So, I suggested Walesa.

Through Doraco I was able to arrange a meeting with Walesa. Though, as it turned out, he couldn't make the Burns Night (he wanted to speak on the Brotherhood of Nations), it was an honour to meet the Nobel Prize Winner. He was very interested in BESO's work and recognised that although he had engineered a political revolution the economic one still lay ahead. I approached him with reverence and admiration but he soon put me at my ease, coming from behind his desk to sit on a couch and chat for half-an-hour. He said he'd have pursued the economic reforms necessary if only the Poles had voted him in a second time! If someone had told me as I was trundling north to Siberia in a Soviet cattle wagon that one day I'd be sitting chatting on a couch with the (former) President of a free and independent Poland, I'd most certainly have thought them mad.

Polish Consulate, Edinburgh. 2nd May 1999. Consul General of Poland, Leszek Wieciech, presents Josef with the Polish Silver Order of Merit.

Through my BESO work I got to know Poland pretty well. This led to people asking me to organise tours around my native land which I've done a dozen or so times, sometimes taking as many as ten or fifteen people.

It was also through BESO that I got to know another volunteer, Margaret Dryden, a former primary head teacher who worked in Cameroon and Kenya helping to set up new schools. Her assignments demanded far more courage and resourcefulness than mine, sometimes involving primitive living conditions and extraordinary diets. Through BESO she became my close friend and then my partner. She encouraged me to continue my work in Poland and elsewhere and often accompanies me to Poland and other places around the world. There is much I have to thank her for and much that I owe her.

I visited Poland for BESO many times until 2004 when they 'retired' me. Since then I have made similar visits as a voluntary consultant invited by individual businesses.

The business consulting I did between 1991 and 2007, which linked Scottish Universities and business firms to Polish ones, was twice recognised by the President of Poland. In May 1999 at a reception at the Polish Consulate in Edinburgh I was awarded the Polish Silver Order of Merit, presented by the Consul General of Poland, Leszek Wieciech. Then, in June 2006, also at a reception at the Polish Consulate in Edinburgh, I received a second award, Officer of the Polish Order of Merit (the equivalent of the British OBE), presented by Consul General Aleksander Dietkow.

Edinburgh, 19th June 2006.
Left: The Polish Consul, A Dietkow, presents Josef with the Officer of the Order of Merit, awarded by the President of Poland.
Right: Josef with Margaret Dryden at the Order of Merit award ceremony.

I am eighty-six years old. I am all the time now losing strength. When I play golf I use the buggy more and more! My mental processes are slowing. I couldn't do the BESO work now but still don't want to give up on Poland. That's why I've adapted once more. As well as occasionally helping specific businesses I've got to know there, I take friends around Poland. Not just casual visitors but people I know who can invest in Poland or help promote my native land in other ways, particularly through marketing.

I still have the inner need to be needed. I can't relax. I have to keep adapting. Life is about creativity. We create all the time.

There is one visit to Poland that I made in 2006 that I must mention: a visit to Szymbark, the only museum outside the old Soviet Union territories dedicated (at least partially) to the Gulag. As well as featuring tourist attractions – such as an upside-down house! – it replicates on a small scale the kind of camp I experienced, complete with watchtowers, barbed wire and a genuine Gulag transport train with a dozen wagons. The similarity with the Nazi death trains is immediately striking, as if they had been designed by the same people.

I was tremendously shocked when I saw the wagons again: all the buried memories re-surfaced. But more than that, this method of transportation continued well after the war when members of the *Armia Krajowa* were still being hunted down along with later opponents of the regime. From September 1939 until the fall of communism some two million Poles were transported in such wagons. It was this tangible reminder of the continuing deportations that really shocked me. They spoke to me and what they said was: had I returned to Poland after the war I would most certainly have been returned to Siberia for the second phase of my 're-education' – my death. They would have made sure that people like me did not survive a second time.

I felt physically sick and had to sit down for half-an-hour to recover, to collect my thoughts, to put my feelings in order. My memory returned to those poor Russians we had left in that camp and, in particular, to my protector Mirsky.

Going inside the wagons was not only to re-live the stinking, primitive conditions; it was to see the images, face after face, of the people who were in the Gulag with me.

I was taken there by Witek Neumann, a businessman I met through BESO, and his wife Kasia. They had family who had been through the Gulag and they took me there because they knew of my experience.

It may be that Szymbark represents a Polish experience which is not really appreciated or understood in the 'West'. It is not on a

The Gulag transport train at Szymbark

scale like Auschwitz, which can disarm even the most hardened observer, but I remember thinking about those French sailors and Fife miners, those naive 'communists' who would not believe our stories, and thinking that even a visit here to Szymbark might give them at least some pause for thought.

I don't really know what Szymbark would mean – or means – to people who never experienced the reality of the Gulag. Maybe not much, though I'd like to think it would mean something. For me, even though there are so many places in Poland commemorating tragedy, it was a very moving experience. But it did raise an ironic smile too: I had survived the Gulag and lived to see the day when I could re-visit it all through a museum – and in a free Poland!

I could not have overcome the difficulties I have met, nor achieved whatever I have, without all the people who have helped, supported and guided me. True communication with true friends has always seen me through. But it is also true that, throughout my life, I have always been guided by instinct or intuition rather than by some lengthy analysis of options and consequences. I have learned to trust that intuition, to accept the outcome of my decisions and not to dwell on what might have been. Whatever happened almost always turned out to be for the best.

There were times I felt I stood on the edge of the abyss but something always prevented me from falling in; something which, at the last moment, compelled me to change direction and save myself. Is this what is called 'destiny'? Was it 'my destiny' that guided the direction of my life and ensured I did not perish? I doubt I shall ever solve that mystery. Maybe it is for the best that I do not know.

In recent years I have often thought of returning to Luck, to close the circle as it were. But I know now I never will. I do not want to. It is no longer my home town. I have no family there now. No friends. And to judge from photographs I've seen, fifty years of Soviet rule did little to preserve its character or charms. It is no longer Polish and I do not want to be a stranger in a strange town. The Luck I grew up in, the Wolyn I explored as a child and a young man, are preserved in my memory, are kept in my heart. Perhaps that, too, is as it should be.

2006. Josef at the Warsaw memorial to Polish prisoners of the Gulag.